EXPLORING THE RHETORIC OF INTERNATIONAL PROFESSIONAL COMMUNICATION:

An Agenda for Teachers and Researchers

Edited by:
Carl R. Lovitt
with Dixie Goswami
Clemson University

Routledge
Taylor & Francis Group
LONDON AND NEW YORK

First published 1999 by Baywood Publishing Company, Inc.

2 Park Square, Milton Park, Abingdon, Oxon OX14 4RN
605 Third Avenue, New York, NY 10017

Routledge is an imprint of the Taylor & Francis Group, an informa business

First issued in paperback 2021

Publisher's Note
The publisher has gone to great lengths to ensure the quality of this reprint but points out that some imperfections in the original copies may be apparent.

Library of Congress Catalog Card Number: 98-44883

Library of Congress Cataloging-in-Publication Data

Exploring the rhetoric of international professional communication :
an agenda for teachers and researchers / edited by Carl R. Lovitt
with Dixie Goswami
p. cm. - - (Baywood's technical communications series)
Includes bibliographical references and index.
ISBN 0-89503-191-4
1. Communication of technical information. I. Lovitt, Carl R., 1946- . II. Goswami, Dixie. III. Series : Baywood's technical communications series (Unnumbered)
T10.5.E97 1998
601' .4 - - dc21 98-44883
CIP

ISBN 13: 978-0-89503-191-4 (hbk)
ISBN 13: 978-1-138-64070-2 (pbk)

Acknowledgments

Assembling this volume has been, at every stage, a collaborative process, and I owe a sincere debt of gratitude to many people who contributed significantly to this cooperative effort. In particular, I would like to recognize the invaluable contributions of three inspiring mentors and special friends: Dixie Goswami, who first urged me to pursue this project and who counseled me wisely throughout the process; H. L. Goodall, who opened my eyes to exciting ways of thinking about corporate ethnography and organizational communication from an international perspective; and Jane Perkins, who never begrudged my incessant interruptions, who either knew the answer to every question or knew where to find it, and whose judgment could always be trusted. I would also like to thank three graduate assistants: Dusti Annan, Nina Lvova, and Myra Whittemore, for shouldering with good humor the unwieldy chore of constructing an index the old-fashioned way; Bobbi Olszewski at Baywood, for her meticulous and attentive production work; Charles Sides, for supporting the project and shepherding it through to completion; and, of course, the volume's distinguished contributors, for responding to the call and for doing world-class work. Finally, an especially warm thank you to Susan Lovitt, who probably wishes I'd forget she used to be a copy-editor and let her get on with the rest of her life.

Table of Contents

PART TWO: Rhetorical Strategies for the Global Workplace

PART THREE: Teaching and Research in International Professional Communication

INTRODUCTION

Rethinking the Role of Culture in International Professional Communication

CARL R. LOVITT

The explosion of interest in international professional communication represents one of the most dramatic developments in the fields of business and technical communication over the past decade. The proliferation of books, special issues of professional journals, scholarly articles, and conference presentations devoted to international professional communication, as well as increased attention to the subject in textbooks, attests to the vitality of research and scholarship in this area of study. Fueling this vitality, however, is a sense of urgency about the need to redress the virtual neglect of this critically important dimension of professional communication. The field of professional communication in this country has abruptly awakened to the limited explanatory power in a global economy of theories and research based exclusively on U.S. models.

Understanding professional communication in a global economy represents a formidable challenge, insofar as it implies nothing less than a wholesale reconceptualization of our discipline. Once we recognize that conceptions of professional communication reflect practices in specific cultural contexts, we can no longer complacently promote insights derived from a single culture as universal, as we have tended to do with U.S.-based models of professional communication. This recognition also calls into question the conventional practice in textbooks of relegating international professional communication to a single chapter, which treats internationalization as a separate topic instead of demonstrating its implications for all aspects of the discipline. As it becomes increasingly likely that our students' professional responsibilities will require them to communicate with audiences from a variety of different cultures, our goal as educators must be to ensure that they possess the knowledge and skills to meet those requirements in any communicative situation.

The challenge of preparing our students to communicate in a global economy has been complicated by the fact that, until recently, relatively little research had been dedicated to this area of study. As a result, the presentation of international topics by textbook authors and educators in professional communication has tended to draw heavily from related fields, such as intercultural communication and international business, which have established a substantial base of research. Relying on research from allied disciplines, however, may have undesirable consequences, especially for a field in the process of redefining itself. In the first place, such imported research may provide answers to questions that are not specifically those of professional communication, thereby obscuring or neglecting areas of inquiry more essential to our discipline. In the second, transplanting research risks imposing the other discipline's conceptions of international professional communication, possibly to the exclusion of conceptions more appropriate for our discipline. In response to such concerns, this volume seeks to define issues that are central to professional communication reconceived as an international discipline.

This volume is particularly concerned with the implications of internationalization for research and teaching in professional communication. Underlying the diversity of assumptions, perspectives, and approaches represented in the volume, questions such as the following provide a unifying refrain: What do we need to understand about communication in the global workplace and in international professional settings and what must we do to acquire that understanding? What knowledge and skills must students possess in order to communicate effectively in a global economy? What pedagogical activities and strategies will help students acquire this knowledge and these skills? Directed primarily at educators and researchers, the authors' responses to these questions constitute an agenda for internationalizing the disciplines of business and technical communication, but their analyses of the global workplace and assessment of concomitant changes in the nature of professional communication have implications for the growing number of individuals whose careers will bring them into contact with members of other cultures.

To create a context for this volume and to clarify what I described above as the pitfalls of importing research from allied disciplines, I begin with a critical examination of the role that has been assigned to cultural analysis in prevailing approaches to international professional communication. My object is not to challenge the essential contributions of cultural analysis to our knowledge and understanding of other cultures, nor is it to deny that cultural factors influence communication. On the contrary, several of the essays in this volume rely heavily on cultural insights for their conclusions. Instead, my intent is to question the appropriateness of privileging cultural analysis as a tool for explaining communication in the global workplace. My concern is that, as a heuristic imported from the fields of anthropology and intercultural communication—neither of which are specifically concerned with the production and management of discourse in professional settings—cultural analysis may be both too blunt an instrument to guide communicators' context-sensitive decisions and too focused on audience analysis to address the complexities of communicating in global workplaces. Intimating some of the limitations of cultural analysis was what led to the inception of this volume. With the intention of prompting other ways of thinking and talking about international professional communication, we conceived of this volume as an opportunity, on the one hand, to showcase research and scholarship that rectified shortcomings inherent in the cultural analysis model and, on the other, to provide a forum for alternative theoretical and practical perspectives on the subject.

THE PROBLEM WITH CULTURAL ANALYSIS

The problem with which I am concerned has resulted from a lack of precision in defining the nature and extent of the relation between culture and international professional communication. Everyone who studies international professional communication agrees that cultural factors are likely to influence how members of that culture communicate. Such a deceptively simple proposition, however, invites at least four questions that have decisive implications for the way we understand and investigate international professional communication:

1. Which cultures influence professional communication?
2. What are the cultural factors that influence professional communication?
3. In what ways do cultural factors influence professional communication?
4. Are cultural factors the only determinants of international professional communication?

Which Cultures Influence Professional Communication?

Considering the variety of cultures to which individuals simultaneously belong, any study of cultural influence entails a critical choice about where to situate the analysis. Thus, when researchers posit a relation between culture and communication, what do they define as the culture that determines professional discursive practices? Is it the culture of organizations and institutions? Of discourse communities? Of localities and regions? Of ethnic groupings and nations? In a move that set a precedent for many subsequent studies of international professional communication, pioneering researchers in the field posited a strong correlation between national cultures and business practices. For example, David Victor, in *International Professional Communication,* the first book devoted exclusively to this subject, states that "[t]he best way for people from two countries to conduct business is to examine the differences and similarities between their nations" [1, p. 3]. Geert Hofstede [2] and Fons Trompenaars [3], whose research into the effects of culture on management has widely influenced studies of international professional communication, similarly organized their findings according to respondents' nationalities. However, as more recent research suggests—such as the studies by Bernhardt and Webb and Keene in this volume—organizational cultures and the cultures of professional discourse communities may have a more decisive influence than national cultures on professional discourse. Thus, a problem with the inaugural decision to examine international professional communication through the lens of national cultures is that it may have distracted researchers from examining sub-cultures that have a more immediate impact on international professional practices.

What are the Cultural Factors that Influence Professional Communication?

Insofar as culture encompasses the vast array of assumptions, values, beliefs, and forms of social behavior and organization that define a particular group, any attempt to analyze the influence of cultural factors on communicative practices must address the challenge of identifying the specific factors that determine the practice under study. However, owing perhaps to the singularly symbiotic relation between culture and communication—such that culture is manifest only through communication and all communication is culturally determined—researchers and educators in the field of international professional communication have consistently espoused the value of acquiring a thorough knowledge of culture. In other words, the tendency has been to sidestep the question of which cultural factors influence professional communication in favor of assuming that all cultural insights have some potential for informing decisions about communication.

There seems to be widespread agreement that the more one knows about a culture, the better able one will be to communicate professionally with members of that culture. David Victor, for example, recommends learning "as much about the other culture as possible" [1, p. 3]. In her recent volume, *International Technical Communication,* Nancy Hoft makes a similar recommendation that technical communicators must consider such cultural variables as politics, economics, society, religion, education, linguistics, and technology in planning documentation for international audiences [4, p. 61]. A similar orientation characterizes presentations of international professional communication in textbooks as well. According to one textbook, "the key guideline is to learn all you can about the other culture and then use that knowledge in your communication" [5, p. 521]; another enjoins students to "find out as much as you can about his or her country's customs, especially the ways in which business is conducted and what is regarded as courteous and discourteous" [6, p. 177]. This is not to suggest that textbook authors do not differentiate among different types of cultural knowledge. Although many textbooks include examples of the "Do's and Taboos" variety, such as oft-repeated warnings not to present the soles of your feet to an Arab or to avoid staring directly into the eyes of Asians, most authors emphasize the importance of going beyond such generalizations. As one author puts it,

> Although these are interesting customs and, because of the etiquette involved, not unimportant, you should research more important questions, such as tradition, religion, history, politics and government, past and present economic conditions, natural resources, and philosophy pertaining to work and business organizations [7, p. 513].

The problem, however, is that, according to Scollon and Scollon, research has shown that only certain cultural factors—notably, "ideology, face systems, forms of discourse, and socialization"—are "of direct significance in discourse between groups" [8, p. 126]. If this is the case, then the injunction to learn as much as possible about cultures may be responsible for again distracting researchers in international professional communication from more relevant objects of study. Instead of deepening our understanding of professional communication, focusing on all aspects of culture risks obscuring essential issues by generating information that may be only marginally relevant to what we are trying to understand.

In What Ways Do Cultural Factors Influence Professional Communication?

In professional communication textbooks, the many international examples about social behaviors that have no immediate bearing on professional communication or business practices clearly reflect the assumption that all cultural knowledge has implications for understanding international professional communication. However, textbooks seldom explain *how* the information is relevant for understanding professional communication. As a result, students may learn how to behave in cultures and improve their ability to interpret cultures, without necessarily understanding how cultural factors affect communication or acquiring the skills to make communicative decisions based on their knowledge of a culture.

As suggested above, the failure to provide such explanations may indicate that the information has no direct implications for communication. Yet, such failures may also point to the larger problem that, as a tool, cultural analysis does not inherently demonstrate how cultural insights translate systematically into implications for communicative practices. The tendency for textbooks to rely on generalizations—such as communications with low context cultures require providing more information or communications with cultures that have vertical structures of authority will tend to be more formal—reflects the difficulty of demonstrating with precision the impact of broad cultural factors on communication. Ironically, the guidelines that textbooks provide for communicating with international audiences are typically not based on insights into any particular culture.[1]

The existence of such problems suggests the need to refine the way that cultural analysis has conventionally been applied to the study of international professional communication. Clearly, greater emphasis needs to be placed on evaluating and demonstrating the relevance of particular cultural factors for understanding professional communication. Two chapters in this volume contribute to this effort by providing models, respectively, for analyzing the impact of culture on documents and for applying cultural content in specific communicative contexts. Through a meticulous analysis of several documents, the chapter by Tebeaux and Driskill illustrates the effect of cultural norms on document design. Similarly, the chapter by Deborah Bosley, which translates cultural insights into guidelines for international visual communication, models how cultural analysis can be specifically applied to professional communication.

Are Cultural Factors the Only Determinants of International Professional Communication?

Positing the acquisition of knowledge about a culture as the condition for understanding international professional communication distorts the relation between culture and professional communication. The goal of learning as much as possible about cultures risks establishing culture as the researcher's object of study, instead of recalling that cultural factors are of interest to professional communicators only to the extent that they illuminate communicative practices. Given the all-encompassing nature of the concept "culture," it may seem paradoxical to imply that cultural analysis may fail to consider essential features of international professional communication. However, in practice, culturally oriented studies of international professional communication rely on models of culture that make no mention of issues that would appear central to international professional communication. Whether the approach calls for examining, on the one hand, technology, religion, education, linguistics, etc. or, on the other, contexting, face saving, uncertainty avoidance, etc., none of these cultural catalogues directly identifies nor specifically anticipates the study of such critically important topics as translation, localization, document design, visual communication,

[1] These guidelines, which do not differ appreciably from one textbook to the next, include the following advice:

- write short simple sentences containing only one idea
- construct short paragraphs that develop one major idea
- use short precise words
- prefer denotative to connotative meanings
- avoid abbreviations, slang, acronyms, technical jargon, sports and military analogies, and other devices that may cause confusion to those unfamiliar with American usage
- become familiar with the traditional format of letters in the target country.

contrastive rhetorics, comparative genre analyses, patterns of reading and processing information, and so on. These omissions may be due in large measure to our having adopted models of cultural analysis that were not developed specifically to study professional communication. Thus, it will be important in the future for professional communicators to target research areas that are essential for understanding international professional communication, whether or not those areas have been anticipated in established models of cultural analysis.

TOWARD A RHETORICAL APPROACH TO INTERNATIONAL PROFESSIONAL COMMUNICATION

Many of the limitations of cultural analysis as a tool for studying international professional communication derive from its inherent emphasis on identifying the *general* factors that influence communication, instead of investigating the specific contexts in which communication occurs. However, any attempt to generalize about the factors that are likely to influence communication in a given culture immediately runs up against the problem that we cannot assume *a priori* that any given cultural factor will necessarily apply in a particular communicative context. As Jorgenson and Steier suggest, "underlying value systems or styles of communication may not even turn out to be the relevant framing features, to the participants, in the process of an ongoing intercultural interaction" [9, p. 74]. Charlene Johnson Forslund, who has studied the communication of health messages to international audiences, issues a similar warning:

> studies in the medical field . . . caution against making generalizations about how an individual from a particular cultural background may react to illness or treatment. Instead, researchers advise health care workers to be aware that a patient's reaction to illness or treatment may be culturally based; thus, health care workers should try to elicit that patient's own individual beliefs about the health care issue being addressed [10, p. 48].

In other words, the more one's analysis is guided by predetermined categories—such as those typically associated with cultural analysis—the less sensitive that analysis will be to the specific features of the situation and the less receptive the researcher will be to data that does not conform to his or her preconceptions. Thus, beyond instructing our students to acquire information about cultures, we need to equip them with more rhetorically sensitive strategies and processes for evaluating cultural interactions and for adapting their communications to the needs of particular situations. Chapters in this volume by Steier, Jorgenson, and Brownell are concerned in particular with adapting communicative strategies to particular rhetorical situations.

The need to develop more rhetorically sensitive models of international professional communication must begin by recognizing that professional communication takes place in professional settings that have their own distinct cultures and that those who participate in international communication often belong to professional discourse communities. This is not to suggest that individuals will not be influenced by their national cultures but rather that they will also—and perhaps even primarily—be shaped by their membership in other discourse systems. For example, Scollon and Scollon note that "cultural differences between people in professional communication are likely to be rather less significant than other differences which arise from being members of different gender or generational discourse systems, or from the conflicts which arise between corporate discourse and professional discourse systems" [8, p. 4]. Chapters in this volume by Bernhardt and Webb and Keene, which respectively examine international communication in the pharmaceutical and aeronautics industries, present similar conclusions about the factors that influence workplace communication. To refine our understanding of international professional communication, research will need to focus increasingly on studying the impact of corporate and discourse community cultures on communication in international professional settings.

If models of international professional communication are to become more rhetorically sensitive, researchers must also recognize the necessity to go beyond an exclusive reliance on cultural insights as a basis for communicating in international professional settings. As Scollon and Scollon remind us,

> when we talk about such large cultural groups we want to avoid the problem of overgeneralization by using the construct "culture" where it does not apply, especially in discussion of discourse in intellectual communication. From an interactional sociolinguistic perspective, discourse is communication between or among individuals. Cultures, however, are large, superordinate categories; they are not individuals. Cultures are a different level of logical analysis from the individual members of cultures. Cultures do not talk to each other; individuals do. In that sense, all communication is interpersonal communication and can never be intercultural communication. "Chinese culture" cannot talk to "Japanese culture" except through the discourse of individual Chinese and individual Japanese people" [8, p. 125].

Professional communication textbooks typically acknowledge this tension between cultural generalizations and individual communications. Most remind students that statements about cultures are based on generalizations (e.g., "the qualities attributed to certain cultures are only generalizations" [11, p. 46]), and they caution students that individuals may not conform to those generalizations (e.g., "be aware of and open to variations and individual differences" [12, p. 69]). The problem is that the general guidelines for communicating with international audiences that we are accustomed to find in textbooks inherently provide no guidance for accommodating these individual variations.

Recognizing that the dynamics of international professional communication are always defined by individual communicative contexts introduces a qualitatively different dimension to this area of study. Effective communication in international professional settings requires supplementing what Jürgen Bolten describes in this volume as "knowledge about a culture" with what he terms "action in an intercultural context," by which he means the strategies and processes that govern interactions between or among members of different cultures. One of the important thrusts of this volume is therefore to elaborate a more process-based model of international professional communication, one that recognizes its object of study as neither static nor monolithic but rather as dynamic and socially constructed.

A DYNAMIC MODEL OF INTERNATIONAL PROFESSIONAL COMMUNICATION

A dynamic model of international professional communication must recognize that culture itself is neither immutable nor impermeable. Even if we could postulate a time when some cultures were relatively impervious to the influence of other cultures, such a myth of cultural purity and stability has become increasingly untenable in our borderless world. The globalization of commerce and of information exchange, coupled with the ease and affordability of international travel, not only provide unprecedented access to perspectives and practices from other cultures but also promote concomitant changes in attitudes and behaviors. The more that members of different cultures come into contact with one another—whether physically or electronically—and the more effort that they devote to preparing for such encounters, the more problematic it becomes to generalize about individual members of a particular culture.

Given that many of the foreign nationals with whom our students may do business will have either studied or worked in the United States, it will become increasingly difficult to anticipate their attitudes and behaviors based on models of their native culture. As Frances Chandler notes in her textbook, "many businesspeople in other countries have become Westernized and no longer exhibit the behaviors normally ascribed to their compatriots" [11, p. 46]. Conversely, what becomes of the convention of opposing the "individualism" of United States culture to the "group orientation" of most Asian cultures, when, as Chandler also notes, "many American businesspeople have come to see the value in the group orientation of other cultures. For example, Japan's group orientation has brought us 'quality circles' " [11, p. 52]?

The very efforts that we, as educators, make to sensitize our students to other cultures will significantly transform their monocultural orientation, making it less likely that their behaviors in international communicative situations will conform to generalizations about our culture. And just as students in the United States are being urged to prepare for intercultural encounters by adjusting their attitudes, expectations, and behaviors, students and professional communicators in other cultures are being similarly educated. The chapter in this volume by

S. Paul Verluyten, which examines the teaching of international professional communication in other countries, offers valuable advice about adjusting pedagogies and course materials to the needs of audiences in different nations.

However, as the following scenario from another textbook suggests, confining the study of international professional communication to the analysis of other cultures could even have the ironic effect of obstructing communication: "You researched your audience's culture, but things aren't going well during your face-to-face meeting. Your audience has also researched your culture. Is it possible you're miscommunicating because you're both too involved in accommodating the other's differences?" [12, p. 111]. Although hypothetical, such a scenario is certainly plausible in light of the increased attention worldwide to preparing students to work in global environments.

What is especially remarkable about such a scenario is that neither participant's actions conform to culturally defined models of behavior: each adapts his or her behavior to the other's anticipated discursive preferences. In other words, this encounter could neither be described nor understood as the juxtaposition of two preexisting cultures; rather, the "culture" that defines this encounter is constructed by the participants during their interaction. Arthur Bell has even suggested that the invention of such hybrid cultures is characteristic of intercultural encounters:

> When you and your own cultural background come into contact with persons of another culture, something new emerges—a middle ground, called a "transaction culture." In this new middle ground, sensitive and often unstated rules and understandings guide behavior. That is, if a member of Culture A interacts with a member of Culture B, neither the cultural rules of A nor those of B are the sole guide for behavior. Instead a mixed set of rules—middle Culture C—develops for the purposes of the interaction.
>
> For example, consider the cultural rules that would guide a business conversation between you and a manager from Japan. You would not speak and act entirely as you would when conversing with American coworkers, nor would the Japanese manager hold fast to Japanese conversational rules and behaviors. Both of you would consciously and subconsciously bend your own cultural habits and assumptions to accommodate the communication needs of the others [13, pp. 452-453].

As Bell makes clear, simply knowing about Cultures A and B will not suffice to explain or to predict what will happen when representatives of those cultures interact. Yet, it is precisely the attempt to base communication on knowledge of the other's culture that accounts for the miscommunication in the scenario discussed earlier. Even though the participants had learned how to analyze each other's culture, their communication breaks down because they lack the necessary skills, tools, and strategies to negotiate a "transaction culture." The chapters in this volume by Bolten and Brownell, which respectively introduce the concepts of the "*inter*culture" and "the third culture," explore the implications of intercultural encounters as contexts that create their own cultures. Steier's chapter is specifically concerned with describing the kinds of process-based "tools" that are necessary to communicate in such constructed spaces.

INTERNATIONAL PROFESSIONAL COMMUNICATION AS SOCIAL CONSTRUCTION

Positing transaction cultures as the loci of international professional communication underscores the importance of investigating the way interactants *produce* discourse in international professional settings. Far from entailing the application of static verities about cultures, international professional communication is constructed by the participants through dialogue, improvisation, and negotiation. Studies in this volume by Bernhardt, Brownell, and Perkins illustrate the extent to which those who participate in international professional communication must invent structures and strategies appropriate to their own contexts. Thus, an important direction for research in this area must be to study the contexts in which such communication occurs and to analyze specific instances of workplace discourse. Such research, however, cannot be relegated to professionals in our field. If our students are to understand that international professional communication is irreducible to a body of knowledge to be mastered, they too must be actively engaged in investigating how communication occurs in the global workplace. Jane Jorgenson's chapter on the student-as-researcher presents a detailed model for engaging students in the study of the international workplace, but most of the other chapters in the volume also incorporate specific recommendations for exposing students to workplace practices. Urging teachers to devise strategies for exposing their students to the international workplace may be the most demanding recommendation of this volume, but the research in this volume presents a compelling case for making the effort.

This volume's call to involve students actively in researching communication in the global workplace has implications for the entire enterprise of teaching professional communication. Workplaces are changing too rapidly for educators to rely exclusively on textbooks and published research to keep their students abreast of current trends and developments. As studies of specific companies by Andrews, Bernhardt, and Perkins in this volume make clear, such factors as the globalization of business, technological advances, corporate restructuring, and new models of collaboration have redefined the nature and landscape of work to such a profound extent that conventional models of operating within stable, monolithic corporate entities can no longer adequately account for workplace practices. It is only by making the involvement of students in workplace research an integral part of their education that we can hope to prepare them for the workplaces they will enter.

To support this call to reorient the study and teaching of international professional communication, this volume seeks therefore to illustrate how changes in corporate structures and practices commit us to rethink the role of the professional communicator. Studies in this volume depict a climate in which communicators must constantly adapt to new modes and genres of communication and to new administrative responsibilities as they navigate a world without borders where someone somewhere is always awake and on the job. Timothy Boswood's chapter methodically examines the implications of transformations in the global workplace for the role of the professional communicator. Chapters by Tim Weiss, Deborah Bosley, and Elizabeth Tebeaux and Linda Driskill specifically illustrate the necessity for professional communicators to make unprecedented decisions involving such complex variables as translation, professional uses of English, visual versus verbal communication, and international document design.

Cultural analysis is indispensable to the study of international professional communication. Understanding cultures increases our awareness of the factors that may influence the way individuals communicate. However, the precedent of privileging a broadly defined model of cultural analysis as the key to understanding international professional communication must give way to a more discerning investigation of the intersections between culture and communication and of the specific ways in which culture affects communication. An important thrust of this volume is therefore to refine the ways in which cultural analysis is applied to the study of international professional communication, notably by demonstrating how specific cultural factors are likely to influence communication. Yet, contributors to the volume also recognize that cultural factors are manifest only in acts of communication and that it is only by examining communication in the contexts in which it occurs that we can identify and understand the specific factors that influence communication.

This volume's call for a more rhetorical conception of international professional communication has far-reaching implications for teaching and research in this area. The need to engage students and faculty in studying communication in the global workplace not only sets a new direction for research in this area but also anticipates fundamental changes in the way we teach professional communication—away from models of information transfer toward models of project-based learning and action research. However, developing rhetorical models of international professional communication also entails a more fundamental reconceptualization of what we are studying in the first place. If we cannot know prior to a communication precisely what factors will determine that communication, then our work as researchers and educators will need increasingly to focus on identifying tools and cultivating strategies 1) for analyzing international communication in context and 2) for constructing transaction cultures that promote effective and equitable communication.

It is not a question of choosing one approach over another. We need to know how cultures may shape communication, but we also need the skills to determine which factors are operative and the skills to devise appropriate forms of communication. To recall Bolten's terms, "knowledge about a culture" and "action in an intercultural context" are both essential if our goals are not only to understand international professional communication but also to prepare our students to function in the global workplace.

REFERENCES

1. D. A. Victor, *International Professional Communication,* HarperCollins, New York, 1992.
2. G. Hofstede, *Culture and Organizations: Software of the Mind,* McGraw-Hill, New York, 1991.
3. F. Trompenaars, *Riding the Waves of Culture. Understanding Diversity in Global Business,* Richard D. Irwin, Chicago, 1994.
4. N. L. Hoft, *International Technical Communication. How to Export Information about High Technology,* John Wiley & Sons, New York, 1995.
5. J. Harcourt, A. C. Krizan, and P. Merrier, *Business Communication,* South-Western, Cincinnati, 1991.
6. P. C. Kolin, *Successful Writing at Work* (4th Edition), Heath, Lexington, Massachusetts, 1994.
7. M. Treece, *Successful Communication for Business and the Professions* (6th Edition), Allyn & Bacon, Needham Heights, Massachusetts, 1994.
8. R. Scollon and S. W. Scollon, *Intercultural Communication* (Language in Society 21), Blackwell, Oxford, United Kingdom, 1995.
9. J. Jorgenson and F. Steier, Social Cybernetic and Constructionist Issues in International Communication, *Teoria Sociologica, II*:3, pp. 63-77, 1994.
10. C. J. Forslund, Analyzing Pictorial Messages Across Cultures, in *International Dimensions of Technical Communication,* D. C. Andrews (ed.), Society for Technical Communication, Arlington, Virginia, 1996.
11. F. G. Chandler, *Fundamentals of Business Communication,* Richard D. Irwin, Chicago, 1995.
12. C. L. Bovée and J. V. Thill, *Business Communication Today* (4th Edition), McGraw-Hill, New York, 1995.
13. A. H. Bell, *Business Communication: Toward 2000,* South-Western, Cincinnati, 1992.

PART ONE:
Professional Communication in the Global Workplace

CHAPTER 1

Communicating in a Global, Multicultural Corporation: Other Metaphors and Strategies

JANE M. PERKINS

Typically professional communicators and professional communication teachers think of international communication in terms of accommodating understanding between distinct language users; even more inclusively, when communication is considered contextually and socially, we think of inter*cultural* communication.[1] Consequently, our research and teaching usually focus on boundaries and obstacles—like check-points at border crossings, or immigration gates, or natural divisions like oceans or mountains, or artificial ones of barb-wire or lines on maps—and we aim to understand differences in languages and cultures emphasized by these separations. Aware of geographic separations and of distinct cultures, we attempt to understand others' languages and habits, to communicate in spite of diversities, to bridge differences, to cross borders, and to transcend boundaries. To those ends we have learned and taught hints for communicating that accommodate non-native speakers and cultural others. Professional communication has and will continue to benefit from "inter" research that helps us understand what confuses and what offends—verbally, visually, proxemically, socially, and so forth. Although this imagery of units (nations, cultures, or

[1] Social perspectives of understanding (for example, [1, 2]) connect language, culture, and knowing.

corporations) with defined borders and boundaries is useful for professional communication, it isn't the only way of thinking.

Today's global and multicultural corporations suggest new metaphors. And, in fact, many who study and write about current corporate changes, and reciprocally influence those changes, are suggesting new metaphors or guiding concepts: Peters, "carnival" [3]; Toffler, "pulsating organization" [4]; Savage, "jazz combos" [5]; and Schein, "harmonies of dissimilar elements or controlled diversity" [6]. These new metaphors emphasize the fluidity and indeterminacy of corporate boundaries, both internally and externally. Especially relevant for international or intercultural communication are announcements from Wriston that "borders are not boundaries" [7, p. 129] or from Ohmae that current corporate management and communication take place in "a borderless world" [8, p. 152]. Professional communication teachers and researchers need to consider additional metaphors, new ways of thinking about international communication; we can begin from an understanding of some current changes in corporations—changes toward borderless, yet multicultural, professional communication.

In the ten years since Faigley introduced a "social perspective" for the study of workplace writing in Odell and Goswami's *Writing in Nonacademic Settings* [2], professional communicators have increasingly understood the importance of studying communication in professional contexts. In particular, that research goal has been supported by ethnographic research of communication in workplace settings. Although we have learned a great deal, for example, about the dynamics of writing and decision-making in an emerging corporation [9], about corporate politics and the structuring of documents [10], about the role of communication in technology transfers [11], about collaborative writing teams [12, 13], and about conflict and inefficiencies in the writing process of an annual report [14], none of these ethnographic studies focuses on international communication, nor do they consider influences of current changes in corporate structures on communication. Professional communication researchers such as Freed [15], Gatien [16], Rogers and Allbritton [17], Sopensky and Modrey [18], and Tebeaux [19] do write about current corporate changes and implications for professional communication; however, their work does not provide in-depth discussion of these corporate and communication changes for international, intercultural, or global communication. Moreover, even if researchers start from assumptions about businesses' growing international markets and, therefore, increased employee communication with international customers, most professional communication researchers/teachers discuss these changes from an "inter" metaphor of boundaries (such as Boiarsky [20]). As a result, most research of international communication continues to offer only hints and suggestions based on language and cultural differences.

Generally, current corporate changes still need to be argued into professional communication to dislodge received assumptions of the mechanized and depersonalized Modern Corporation, which dominates teaching of most professional communication courses. A few researchers, however, are beginning to connect corporate changes with professional communication pedagogy, and in the process, they are supplanting the boundary metaphor of reified borders and static homogeneous cultures. This chapter is not the place for me to argue out the substantive impact of current corporate changes or the lack thereof—of the ways in which management theories, technology, and communication interact and affect societies at large or balances of social and political power or the end results on social structures or the environment. I'm not suggesting that we can separate our professional communication teaching or our research from issues such as these; much valuable writing has argued for this kind of ethical questioning and responsibility in our work (see, for example, Jacobi [21], Katz [22], Miller [23], Perkins [24]). Neither, however, can we ignore specific workplace contexts.

Beginning with "a world of interlinked economies and communication networks" [25, p. 407], Weiss writes about the increasing important roles of professional communicators as translators in a "Borderless World" [25, p. 407]. Based on his personal international experiences and on interviews with Moroccan professors, Weiss recognizes the changing needs of professional communication students in a "more complex world of international information exchange" [25, p. 419]. His research in translation and my ethnographic research of a global, multicultural spin-off software development corporation (VisionCorps)[2] both emphasize new approaches for teaching and working from a global communication perspective. We both believe the boundary metaphor has the potential of reinforcing cultural stereotypes. Alternatively, Weiss suggests focusing on the individual; I argue for an understanding of "culture" as fluid, layered, overlapping—distinguishable not by boundaries of inclusion and exclusion, but by prototypical features. Weiss concludes that we need

> to address the individual in international contexts, rather than the categorical Chinese, Japanese, or Arab audience. The generalizations and stereotypes about cultures are good to know, but finally it is the individual whom we address and that individual abroad will be just as complex and unpredictable as any individual in our own country [25, p. 422].

I, too, attempt to dislodge the boundary metaphor with my ethnographic stories about the global and multicultural communication of VisionCorps. But first, I want to add to the mix of this discussion some relevant ideas of current corporate theorists and of interpretive ethnographers. Do not read the following section as a theoretical framework of direct correspondence to claims I make of VisionCorps' global, multicultural communication. Rather, think of these theorists' discussions as included within, and yet at the same time comprising, a

[2] VisionCorps is a pseudonym for the software development corporation that was the site of my more than three-year ethnographic research; the names of employees are also pseudonyms, but the stories are of actual people, places, and happenings.

cultural milieu—as anthropologists say—or as within the heteroglossia in which we teach and research, perhaps a heteroglossia of global scale.[3] Peters [3], Drucker [26, 27], Ohmae [8], and Wriston [7] explain that global influences—and therefore global communication—are integral to current corporate changes in management theories, organizational structuring, and technology. Many anthropologists, including Dubinskas [28] and Rosaldo [29, 30], also suggest important theoretical perspectives of multicultural communication. All of these theorists problematize concepts of static cultures of determinate boundaries and cultural stereotypes.

THEORISTS OF CORPORATE CHANGE AND CULTURAL KNOWING

Corporate theorists, such as Peters [3] and Drucker [26, 27], write about new global markets and increased global marketing competition as integral to current corporate changes. Peters gathers bits and pieces to tell stories of corporate innovations; many of his success stories are of international corporations, more specifically, of multinationals. A global perspective is inherent in Peters' proselytizing; for example, he predicts that "only a fickle, decentralized operation will survive in a fickle, decentralized global economy" [3, p. 9]. He cites for support the shift in information technologies; although originally geared primarily for United States' markets, they now have markets that are 80 percent non-U.S. [3, p. 271].[4] Drucker glosses in more prophetic overviews the causes and implications of a global economy: "Money going transnational outflanks the nation-state by nullifying national economic policy. Information going transnational outflanks the nation-state by undermining (in fact, destroying) the identification of 'national' with 'cultural' identity" [27, p. 145].

Besides explaining causes of the globalization that has and is occurring, both Peters and Drucker also write about what could be seen as a paradox of globalization; they argue that, although people and cultures are in closer contact than ever before, cultural assimilation—a homogenizing consequence—is not occurring. Peters contends that the homogenization of differences is exactly counter to current corporate endeavors; diversity is necessary and, therefore, encouraged for the kind of innovative thinking and creative problem solving that must fuel his vision of "liberation management" [3, p. 414]. Peters summarizes: "The message . . . is fashion. Translation: diversity. The globe is shrinking. But as it shrinks, oddly enough, it is becoming more diverse, not less—no matter where you look"

[3] Rather than dissecting an argument into discrete parts or building informational blocks toward an unequivocal contention, my approach is more that of button soup; understanding develops from a blend of ideas, which although perhaps identifiable in the end product, offers new aromas and flavors.

[4] VisionCorps falls into this corporate category; their business is designing software tools for information management and their markets are international in scope.

[3, p. 667]. Drucker too believes that globalization, while de-emphasizing national borders and cultures, does not result in monoculturalism or the disappearance of unique and valued cultural differences. Rather than attributing diversity of cultures to Peters' "fashionizing"—the ability to create, produce, and market products for diverse personal and idiosyncratic tastes, Drucker explains that bigness, in a unified sense, is no longer an advantage, especially in political or corporate units. Furthermore, he predicts that "the more transnational the world becomes, the more tribal it will also be" [27, p. 155] because of the need for roots and community. "A system in which transnational, regional, nation-state, and local, even tribal, structures compete and co-exist . . . ha[s] already happened"; therefore, Drucker believes his purpose is to describe these multiple cultural layers [27, p. 4]. As they describe current corporate changes, Peters and Drucker establish a global perspective for conducting business as integral to those changes. For them, globalization does not just dissolve national borders in exchange for large-scale uniformity. Their global perspective means that the metaphor of borders, especially national borders, is clumsy and inappropriate in a fast-paced business world aiming for diverse and rapidly changing market segments. They see markets and cultural structures segmented in new and changing ways.

Similar to Peters and Drucker, Ohmae, an international management consultant who heads McKinsey's office in Tokyo, agrees about national boundaries and about the fine granularity of global market segmentation:

> On a political map, the boundaries between countries are as clear as ever. But on a competitive map, a map showing the real flows of financial and industrial activity, those boundaries have largely disappeared. What has eaten them away is the persistent, ever speedier flow of information . . . [8, p. 153].

Rather than thinking of international markets in terms of national divisions, "our German market," "our Brazilian market," or "our Korean market,"—each a national entity of language and cultural stereotype—a global perspective means that while "some top-of-the-line tastes and preferences have become common across the Triad" [8, p. 159], other markets are unique and local within discrete segments of a nation. Ohmae explains that "when it comes to product strategy, managing in a borderless world doesn't mean managing by averages" [8, p. 155]; it "means building the capability to understand and respond to customer needs and business system requirements in each critical market" [8, p. 159]. In addition to this picture of globalization with ever-shifting, niche markets, Ohmae overlays other dimensions. He draws on his personal experiences to support the existence of a rapidly growing, global, English business language. Business English isn't eliminating languages but is learned in addition to others and is used to facilitate global communication. The boundary metaphor is further supplanted by another dimension of globalization. Ohmae describes global high-tech contexts of knowledge where "scientists can move from one laboratory to another and start working the next day with few hesitations or problems" [8, p. 154]. Enhanced by technological innovations, specialists are able to share information with other specialists dispersed around the globe, linking these specialists in new ways and creating new, sophisticated alliances.

These global interactions between specialists, linked to share information and to build knowledge, is just one variety of the many kinds of alliances discussed by Wriston:

> On the global business front, the new word is "alliances." Almost every day one reads about an American company and a Japanese company or a Swedish company and a German company—the list of combinations is endless—forming an alliance to offer a new product in a multinational horizontal integration of manufacturing, marketing, finance, and research. As these alliances grow and strengthen over time, it will become harder and harder for politicians to unscramble the emerging global economy and reassert their declining power to regulate national life [7, p. 10].

With a focus on the international, Wriston explains another feature of current corporate structural changes: new and creative mergings of expertise or core competencies. When corporations move toward decentralizing, giving up efforts to control all aspects of production in exchange for a leaner and speedier approach toward research and production processes, corporations fill voids by developing innovative alliances with other corporations. In simple terms, the big corporation no longer does it all. These alliances can enhance all or any part of a corporation from development activities to customer education.[5] Therefore, much international communication is no longer about separations between "us" The Producer (of goods or services) and "them" The Customer (or consumer of those goods or services); this sharp distinction is blurred in many ways and often the "them" is also a part of "us." "The traditional multinational economy in which 'products' are exported" Wriston argues, "is being replaced by a truly global one in which value is added in several countries" [7, p. 77].

Similar to the way corporate theorists are arguing that traditional, national boundaries are less significant for today's global business, anthropologists—those whose expertise and disciplinary history comprise studying cultures—are questioning the categorizing or grouping of peoples as cultures according to national distinctions, or for that matter, according to any seemingly fixed borders. Cultures of study are arbitrary—not "discovered" intact. Just as Ohmae says that being a global corporation does not mean thinking in terms of "averages," many anthropologists' aims have shifted from understanding cultural generalizations to creating awareness of the indeterminacy of cultural boundaries, the layers of cultural belongings, and the relationships of power inherent in "identifying" and positioning oneself in relationship to other cultures. Therefore, the value of

[5] VisionCorps is a maze of alliances; I'll describe some of them to you in the next section and explain how these alliances affect communication.

bringing interpretive anthropological theory[6] into the discussion of international communication is twofold: 1) much self-reflexive ethnographic writing addresses issues of cultural boundaries and 2) anthropologists understand the difficulties in studying "the other," especially the hard-to-avoid dangers of generalizing or stereotyping that can lead to comparisons of "us the knowing" in contrast to "them the less-knowing."

In his ethnography of a high-tech organization, "Cultural Constructs: The Many Faces of Time," Dubinskas writes about the socially-constructed nature of time as a defining feature of culture, rather than describing more traditional categorizations. He argues both that "alternative, multiple views of time" [28, p. 9] exist within traditional cultural boundaries (a nation or a corporation), and that views of time also span geographic cultures, linking or creating new professional cultures or "subcultures." In other words, scientists located around the world who work in a common high-tech field may have a more similar perspective of time than a scientist might share with co-workers in the same geographic organization. Which then is the culture? The geographic organization or the trans-geographic community with specialized bonds? Culture means more than a common geography or common language; it can be as nebulous as the sharing of professional interests. And cultural membership can be infinitely multiple, layered, intersecting—disorderly. This disorder is also an important aspect of global communication.

As we increasingly embrace global communication issues, professional communication teachers and researchers can perhaps benefit most from anthropologists' theories of cultural others. In particular, we can learn from what Rosaldo claims is "a creative crisis of reorientation and renewal" [29, p. 28]. Therefore, rather than experience a similarly haunting history and treacherous learning curve, we can benefit from valuable theorizing about intercultural knowing and consider, in particular, our relationships to cultural others in the global workplace. As he argues for experimental ethnographies, Rosaldo exposes the colonialist dangers of the "classic norms" of anthropology:

> Although the classic vision of unique cultural patterns has proven merit, it also has serious limitations. It emphasizes shared patterns at the expense of processes of change and internal inconsistencies, conflicts, and contradictions. By defining culture as a set of shared meanings, classic norms of analysis make it difficult to study zones of difference within and between cultures [29, p. 28].

To emphasize the limitations of the classic perspective in which "each cultural pattern appears as unique and self-contained as each design in a kaleidoscope"

[6] Because anthropology like other disciplines is theoretically non-unified, it is difficult to label "schools of thought"; interpretive, postmodern, and experimental are some of the descriptors used to differentiate this work and theorizing from traditional, classic, realist anthropology.

[29, p. 27], Rosaldo tells the story of the Lone Ethnographer. He describes classic ethnography as being "dominated by the concepts of structure, codes, and norms" . . . which developed a distanced, objective "mode of writing that normalized life by describing social activities as if they were always repeated in the same manner by everyone in the group" [29, p. 42].

In their reductive unifications, classic ethnographies depict cultures as unchanging and homogeneous. However, the stilted, overgeneralized, and limited cultural views aren't the main problem. The real danger exists in a simplistic "us"—"them" attitude that results:

> The Lone Ethnographer depicted the colonialized as members of a harmonious, internally homogeneous, unchanging culture. When so described, the culture appeared to "need" progress, or economic and moral uplifting. In addition, the "timeless traditional culture" served as a self-congratulatory reference point against which Western civilization could measure its own progressive historical evolution. The civilizing journey was conceived more as a rise than a fall, a process more of elevation than degradation (a long, arduous journey upward, culminating in "us") [29, p. 31].

These dangers of reductive unifications, which lead to cultural stereotyping, lurk in the prevailing boundary metaphors used by international communicators, teachers, and researchers. Ironically, while we have attempted to enlighten ethnocentric attitudes of "everyone thinks and acts like I do," for which survival hints have been gathered in efforts to avoid cultural faux pas, we have often reinforced reductionist generalizations, resulting in an "us"—"them" attitude.

An alternative may be to foster a global/multicultural, rather than an [inter]national, perspective. Corporate theorists recommend a global perspective because it seems appropriate for worldwide workplace changes and because it seems to be the approach of successful businesspeople; interpretive ethnographers embrace a multicultural perspective because it offers alternatives to the classic norm of overgeneralizing and, therefore, colonializing. Ohmae suggests "building a new value system that emphasizes seeing and thinking globally" [8, p. 153]. This value system depends on a repositioning of corporations in terms of customers: "the corporation sees itself as equidistant from all its key customers" [8, p. 153]; in this value system, there are no borders, no gaps, no overriding "us"—"them." Proclaiming that "the Lone Ethnographer's guiding fiction of cultural compartments has crumbled . . ." [29, p. 45], Rosaldo recommends an "agenda of theory." This agenda depends not on "discrete variables and lawlike generalizations" but rather on "the interplay of different factors as they unfold within specific cases" [29, p. 34].

In addition to the already frustrating challenge of teaching numerous, cultural conventions, these discussions by corporate theorists and interpretive anthropologists further complicate and refocus pedagogy of international professional communication. But they also suggest new pedagogical directions. If we are persuaded by these corporate theorists and cultural anthropologists, we realize that staid overgeneralizations of national characteristics don't get us very far in effectively communicating. The approach is problematic. We need to study and teach a variety of cultural theories and to research specific workplace contexts, in order to

- understand the variety and pervasiveness of local contexts rather than the overgeneralized and stereotypical
- emphasize individuals and the disarray of cultures rather than to make still lifes for easy classification or codification
- tell and retell numerous small stories of communication—not of boundaries but of interplay and repertoire building

Professional communicators need ethnographic studies of global, multicultural workplaces. Based on this research, teachers of professional communication will gain understanding of communication in global, workplace environments and develop relevant teaching theories and practices.

STORIES OF A GLOBAL WORKPLACE

For three and a half years I studied the communication of VisionCorps, in particular the communication processes of the three owners and approximately fifty employees, and the ways their documents (hardcopy and electronic) were both a result of the new corporate structure and were also concurrently "writing" this spin-off corporation. For those years of fieldwork during most of which I was also a part-time documentation specialist, VisionCorps' global, multicultural status permeated the corporate culture(s). Therefore, from my formal and ongoing interviews/conversations and observations, and from multitudes of casual interactions and involvements, I have stories to tell of cultural members and their global communications. These aren't the first stories I've told of VisionCorps; however, this telling is the one most focused on global professional communication. Part of the rich challenge, but also the frustration, of being an interpretive ethnographer stems from a belief that ethnographers identify the culture or cultures of study from among layers of cultural memberships, as they also identify the perspective or focus from which they tell their stories. Interpretive ethnographers' research is not about discovering existing cultural truths; they make contextual meaning, in partnership with the cultural members.

The following categorizing results from my analysis of the culture(s) and adds to the meaning-making of my interpretations. In the final section of this chapter, I will discuss the effect my research has had on my pedagogical reflections and subsequent innovations for my professional communication courses. In my ethnographic research of VisionCorps, I have learned a great deal; the following ideas are particularly significant for rethinking international professional communication with its metaphors of boundaries:

- VisionCorps was born global
- VisionCorps[7] staff and their many customers and partners generally communicate in one language—technical English
- VisionCorps owners and employees communicate globally with independent authority
- VisionCorps cultural members are theorizing and creating a global perspective

Born Global

Located in the landlocked Midwest and operating with a pared-down, lean staff, carefully chosen from the original Parent Corp research and development division, the owners and employees of this not untypical corporation conduct business on a global scale with little or no foresight or formal training devoted to effective international professional communication. VisionCorps, like many other young businesses in the information industry, was "born global" [13, p. 26]. As part of Parent Corp, one of the world's largest computer companies, the owners and the majority of the employees who now comprise VisionCorps developed highly-specialized software. The users for their software are not the general public. Although the software products vary in sophistication, many users are experienced system administrators, who manage, for example, the information for running a nations' banking operations or railroad system. When downsizing loomed and a rumored closing of the midwest R & D and production facility spread, a senior manager of Parent Corp suggested the spin-off. VisionCorps came into existence with contractual agreements to Parent Corp and existing customers around the world. For VisionCorps, expanding into international markets was never an issue; the issue was supporting their products for their existing customers and growing new products and new customers.

For VisionCorps to be born global and to survive in the volatile software industry, the corporation not only needed established customers around the world, but the staff had to be able to sustain those customer-relationships. Ben, the company president, explained VisionCorps' global, spin-off connections, stressing the differences between the way Parent Corp conducted its international business and VisionCorps' aim to establish global networks. He had just returned from a state-sponsored trade mission to Japan and was eager to talk about VisionCorps' global communication—especially the lure and opportunities of that "huge dense market" for VisionCorps. "I've been working with some of these Japanese business men for 20 years, and I got nowhere." However, as he and

[7] By *not* using an apostrophe to refer to VisionCorps' staff, employees, etc., I avoid saying that these people are employees *of* the corporation, with all of the Modernist concepts of corporate objectivity and possession that entails. Rather, I am saying that these people comprise the corporation.

others in VisionCorps begin conducting business with increasing global awareness, Ben believes they will be able to change from this former stagnate approach. He's optimistic because of VisionCorps owners' and employees' better customer understanding, which depends on an innovative corporate structure and the related benefit of improved timing or rapid response. While Parent Corp focuses on developing new products and selling them, VisionCorps tries to focus on the customer from the beginning. As a small, flexible R & D company, its goal is to develop new features and products that customers request, to involve those customers in design stages by often arranging for them to work onsite alongside VisionCorps employees, and to include those customers and others in refining and testing products as beta test sites. But that's just a hint of VisionCorps' elaborate networks, which are supported by their corporate structure. In addition to alliances with customers that complicate and recreate business relationships, VisionCorps' supplier network is also blurred: they are both a supplier for Parent Corp and a competitor with the same products; they have numerous "partnerships" with other small software developers who have unique expertise or value joining forces to bundle products rather than offering them competitively; they work with corporate trainers to fine-tune products that the trainers offer to customers in seminars or workshops, creating a need for VisionCorps' products and those of other small companies; and through a variety of agreements, they are experimenting with engaging distributors based in other countries, who are also sometimes customers, to promote and sell their software. These alliances, partnerships, and interactions occur on a global level, usually without much thought to languages, cultures, or governments. VisionCorps marketing specialists claim that they spend at least 80 percent of their time with international customers and alliances.

VisionCorps' networks of connections with "customers" and "suppliers" is developed and maintained because of its corporate structure. To facilitate these many interactions, almost all VisionCorps employees are involved in generally informal customer communications. While some communication is more formal and ritualized, such as that between Parent Corp executives and VisionCorps' three owners, the majority of "external" communication is casual, on an as-needed basis to conduct day-to-day business. For example, Paul, a highly-regarded veteran software engineer, supports one of the company's most sophisticated products for highly-specialized users. During one of our frequent meetings to refine cautionary statements for the product's documentation, he responded to my questions about the users of the product, "I can probably count the number of users on my two hands, and have talked with on the phone or have met most of them; if they're confused, they'll call me, and I'll walk them through any problems." This kind of direct customer interaction, early and often in the development process, is an ideal that motivates many developers and legitimizes the fast-paced and chaotic communication processes. Those customer inquiries, by the way, come from all over the world and usually by phone, sometimes by e-mail or fax. In the rigid hierarchy that Ben describes as Parent Corp, few employees, usually those in sales, support, or installation, have direct contact with customers. In contrast, VisionCorps employees' interactions with customers and other business alliances are continual. Because so many of those connections and communications are with people from around the world, a global perspective pervades this small corporate environment.

Communicating in Technical English

I have a vivid memory of sitting in a Documentation Team meeting[8] during the second month of my fieldwork when a developer interrupted to ask an urgent question about a soon-to-be-released product. Uncomfortable with her interruption, she hurried through a three-minute discussion with the senior documentation specialist. Except for a few short verbs, all the words spoken were acronyms and a few technical terms—their short-cut and precise language of specialized computer functions, feature components, and product names. This typical occurrence helped me realize that, because of my assumptions and grounding in traditional approaches to international professional communication, I was asking a lot of wrong questions, especially those about accommodating language and cultural differences. Because of the acronistic and technical meta-language in which VisionCorps staff, customers, and partners most often communicate, divisions or boundaries caused by language and cultural differences are not normal concerns. And although communication occurs among people from many countries, it is frequently interpersonal communication, between people who have names, faces, and ongoing work relationships. A number of VisionCorps employees stressed that it was a good thing they could communicate in their technical English; the sheer number of countries (England, France, Sweden, Italy, Japan, Turkey, Saudi Arabia, Mexico, Spain, Portugal, Australia, Korea, Taiwan, and Brazil) where their customers and partners are located make communicating in any other language a logistic impossibility.

VisionCorps employees communicate in technical English via the array of conventional business communication mediums, whatever it takes to facilitate their work. Individuals' communication styles, their current projects, and sometimes technical constraints determine whether they reach for the phone, access their e-mail, or do both at the same time. And they often communicate face-to-face with customers who visit onsite (and sometimes in less technical English when the visits are for extended lengths of time and casual relationships develop that continue out of the office), for international conference presentations and trade shows, with participants in classes taught by developers, and during marketing

[8] VisionCorps staff often experiment with team configurations; owners and employees are members of many teams concurrently. For example, documentation specialists participate as part of a group to make writing decisions and as members of cross-functional project teams, usually averaging two or more.

trips. Letters from potential customers, which follow as a result of trade shows, are also primarily in English, both as faxes and hard copy. On a rare occasion, correspondence must be translated by the local Spanish teacher or the senior documentation specialist who lived in France. Most global communications, however, are by telephone and e-mail. Discussions about technical information are usually conducted by e-mail, between VisionCorps specialists and specialists anywhere in the world, in short messages of technical English. Marketing specialists often work at odd hours of the night to place international phone calls. And global telephone communication is important for Jay, the employee responsible for registering trademarks in all countries where products are sold and for negotiating many contract details. Sometimes he communicates with patent offices in small communities where the person does not speak much English or have familiarity with computer terminology. Jay explained he has developed "overall strategies for working internationally, rather than country-specific approaches." In explanation, Jay and other VisionCorps employees claimed that their global communication attitudes and strategies have been shaped, in large part, through their day-to-day work with VisionCorps computer engineers who have international origins.

A global, multicultural presence, therefore, is part of VisionCorps' makeup. Communicating in primarily technical English is as natural for overseas telephone conversations and e-mail as it is to discussing new product features with Krishna, Inh, Anil, or any of the interns from the university's computer engineering program who rotate through VisionCorps. These developers are as highly regarded as any employees in the corporation: Krishna as a long-time and integral member of the most innovative development team; Inh as the computer engineer who supports and designs new features for the product that results in more than 50 percent of the corporation's total revenue; and Anil as the new Unix wizard, lured to the group to develop additional products for this computer platform. A blend of accents—from Viet Nam and India on a daily basis, and frequently from other customers' and partners' locals—is the norm; everyone is accustomed to communicating to make their ideas understood and to learning with each other, using the continually evolving technical English of their fast-paced, high-tech industry.

Communicating Globally with Independent Authority

Because of the non-hierarchical organizational structure of the corporation and the often chaotic pace of daily work, VisionCorps owners and employees communicate globally with the same kind of independent authority that guides all their communications. This communicative openness differs from what most of these employees experienced with Parent Corp: 1) the employees would not even be communicating with customers and partners, especially globally, on a day-to-day basis and 2) VisionCorps tends to avoid the formal document review cycles and routing of information that were essential in the Parent Corp structure. VisionCorps owners and employees are willing to trade control and standardized safeguards for speed and flexibility. They justify this trade off because they believe their competitive advantage offers quick reaction time and direct customer interactions, and because they literally comprise a select group of employees,[9] highly-regarded experts in their areas.

In general, the lack of formal checks on communications does not mean that communication is less valued; it means rather that colleagues must be trusted to know their strengths and limitations, and to develop informal methods for feedback and edits. Documents intended for wide audiences—including global readers, such as conference papers, corporate newsletters, and advertising material, are circulated inhouse, especially through the documentation specialists and others with relevant technical expertise. Likewise, important letters to customers might be casually circulated. Generally, individuals with strong writing abilities gravitate toward positions in which those abilities are needed, or more commonly for VisionCorps, individuals invent positions for themselves as they identify opportunities. Within the first year of the corporation, two former software developers created positions, in marketing and in liaisons/patent rights, both of which require excellent global communication abilities. However, communications don't always work smoothly.

While employees often seek each other's advice on their written documents or float questions over the cubical walls, for example about subject/verb agreement, they also instill communication standards through inhouse broadcast e-mail, with its company-wide access. A cultural expectation of well-written e-mail, with standard spelling and grammar, is crudely enforced by cutting jibes to those who err. As a result, the upgraded communication software with spell check was a jubilant event, and still some employees shy away from posting messages, thus avoiding any potential negative reactions. These e-mail standards extend to external communication, especially since customers and partners are often included in some of these "inhouse" project groups.

In such an open communication environment, another area of concern is proprietary information. When the same contact can be both your customer and your competition, or when employees can quickly and easily bounce information around the world on e-mail, the unintentional sharing of proprietary information is less a matter of distrust than of carelessness. Again, informal safeguards have evolved to reinforce the significance of communications. Although most process documents that guide projects and thus everyone's activities are circulated and updated by all employees electronically, some highly

[9] When the spin-off was formed, the three owners agonized over their offers to fewer than 50 of the then over 100 employees in the midwest R & D division.

proprietary documents, for example the VisionCorps Strategy Document in which the direction of the corporation is outlined, are produced only in limited copies. Some documents that show up in employees' mailboxes, both physical and electronic, are emblazoned "confidential," and most often employees remind each other verbally.

While most VisionCorps employees communicate globally with independent authority, the documentation specialists are somewhat of an exception. Because Parent Corp is still their major buyer, documentation, in general, must adhere to Parent Corp's more rigid and layered review and editing cycles and to their standards for documenting software products. These standards are not influenced by any analyses of global users; the assumption behind the documentation is that users are all speakers and cultural members of American English. Therefore, the documentation specialists have never approached writing manuals with globalization concepts nor have they been encouraged to communicate with global users from a documentation usability perspective. Because she has an extensive global background, teaching and writing in Europe, Africa, and Saudi Arabia, the senior documentation specialist is especially frustrated by the lack of cultural awareness in the standards for the software manuals. Sandy believes that "all of their products need to be written with cultural awareness, and that as documentation becomes more of an online feature, the demand is even greater." While documentation specialists are increasingly caught between Parent Corp's standards and processes and VisionCorps' demands for more flexible and shorter development cycles, the pressures of change may also usher in more global awareness in documentation.

Theorizing and Creating a Global Perspective

A metaphor of boundaried spaces isn't applicable to VisionCorps' corporate structure, either for internal divisions of work or external relationships with customers/suppliers; it also has little value for describing or guiding owners' and employees' communication. No easy assumptions, however, are replacing that received notion, which has long-dominated world views with clear-cut, easy separations of ours and theirs, of us and them. On a micro-level, VisionCorps owners and employees search for answers to the same issues of language assimilation and homogeneous dominance in tension with the advantages and valuing of unique differences. Much of their ongoing discussion focuses on the internationalization (I18N)[10] of their software products.

[10]As with most other terms in the fast-paced computer industry, internationalization has been shortened, in this case, only for written use. The eighteen letters making up the middle of the term are replaced with the number and only the beginning and ending letters remain.

As with many other corporate issues, the owners and many of the actively involved and vocal employees are divided, between marketing and R & D focuses, in their opinions about internationalization, defined by O'Donnell as "the process of generalizing computer systems so that they can handle a variety of linguistic and cultural conventions" [32, p. 418]. The president and the marketing specialists are gambling that internationalization and its counterpart, localization, "The process of providing language- or culture-specific information for computer systems" [32, p. 419], will give them an advantage with their wordwide customers. The owner, who leads R & D efforts, and many of the computer engineers are more skeptical of the process, one that involves rewriting all of the source code for the company's software products. They believe it would just take too much valuable time away from their work on new features and products, especially when they believe they already know their customers—users who have been generally pleased with the current American-English versions. Therefore, they are entering the I18N process cautiously.

One of VisionCorps most creative computer engineers, who's usually involved in leading-edge product development, has been selected to begin internationalizing the source code for an old stand-by product. Although he isn't convinced of the value of the task because he doesn't think current users will benefit much, Tony thinks internationalized and localized software might help them open some new markets and also prepare them if "some foreign government suddenly gets restrictive about protecting their language." Tony believes his project will stretch over many months or even years since he works primarily by himself, except for his localization counterpart in Japan. The two e-mail frequently to coordinate their processes, and Yoshimi came to work in VisionCorps when the two computer engineers most needed intense face-to-face collaboration in the early project stages. When it comes time to field-test the internationalized product, Tony will go onsite in Japan. From Tony's perspective, he and Yoshimi have no language or cultural problems to overcome; they communicate easily: "I know Yoshimi pretty well now. We don't really spend time thinking about communicating." While Tony is an adept communicator, he began this project by turning to resources to help him understand more about communication and cultural differences. O'Donnell's *Programming for the World* helped him recognize cultural assumptions, especially those that affect the way engineers write code: "Assumptions like: *A—Z includes every letter.* Or: *spaces separate words.* Or how about: *'characters' and 'bytes' are the same thing*" [32, p. 9]. Depending on how it is applied, internationalization can be a means to manage cultural specifics, such as differences in the way dates are ordered; or taken to an extreme, it can be an attempt to sanitize language, to eliminate differences. On a larger scale than this specific project, Tony is concurrently analyzing and writing guidelines for others who will be internationalizing code in the future. And thus, in his everyday work, he balances complexities of global communication, issues of promoting assimilation and of maintaining difference.

TEACHING GLOBAL COMMUNICATION

Stories of VisionCorps' global workplace communication and theories of corporate change and of interpretive ethnography combine, interactively, to suggest students' needs, which guide our design of professional communication courses. Most importantly these stories and theories add to the collapse of an overriding metaphor of boundaries that separates and rigidly problematizes communication between people of different nations, too often reducing global communication and its study to admonitions based on reductive stereotypes. Alternatively, stories and theories such as these address students' global professional communication needs and suggest specific practices for teaching.[11]

Pervasiveness of Global Professional Communication

Students need to be aware of the pervasiveness of global professional communication. Although students may hear that it is a global world, they often aren't aware of the significance of that statement for their own futures. For example, many people would not realize that small, midwestern VisionCorps does 80 percent of its business with customers and partners located out of the United States. When students have assignments that take them out of the classroom and into the workplace, asking questions of people in jobs that they might soon be filling, they become more aware of the pervasiveness of global business and communication. Ideally, students would work on an assignment that requires them to negotiate an interview with a person in their field and to visit that person in the workplace; an alternative or supplement might be telephone or e-mail interviews. At the same time they learn and practice interviewing skills, students produce a number of written documents, and realize the importance of professional networking.

Professional Communication in the Workplace

Students need to understand how professionals communicate in the workplace, especially globally. Not only do students need to know the extent of global communication, they need to learn specifics about how that communication is accomplished. As students develop their interview opportunities, they can formulate questions that will inform their understanding of global professional communication. A class might then maximize this learning by sharing their information in class presentations, in discussion, and/or as part of a written group project. They might learn answers to questions such as the following: What methods of communication are used in their future profession? And to what extent?

[11]These objectives and practices for teaching are appropriate for both general courses in professional communication, such as Technical Communication and Business Communication, and courses specifically focused on International Professional Communication.

Face-to-face and verbally? Written in hardcopy and e-mail? What formats? What kinds of process documents guide the organization's work? How do workplace teams communicate, especially if they are comprised of people in diverse locations? What constraints impact global communication decisions? What role do visuals play in global professional communication? Students can ask these questions and others during their main interview assignment or, perhaps, as part of chats and discussions they initiate with e-mail and WWW respondents. A class might also design Web projects, such as establishing a site to manage information about global communication, and confront complex, global audience decisions.

Analysis of Specific Global Communications

Students need to analyze specific global communications and to develop strategies for producing them effectively. During all of their information generating assignments, students will also want to obtain copies of professional documents, especially those produced by or for global writers and readers. Teachers can add some of their classic examples and help students learn methods of analysis and evaluation, which students can then practice as they write reports or present their analyses orally, much like workplace activities. Rather than teaching stock "rules," this method encourages rhetorical and contextual analysis. Teachers may also ask students to plan for and to develop methods for feedback and editing of documents or oral presentations. Students might be required to present a proposal to a room of international students and afterwards lead a discussion to elicit reactions and suggestions from their listeners. Or students might design an instruction manual for global users and conduct a series of iterative usability tests.

Theories of Global Communication

Students need to develop theories of global communication. As an ongoing part of the course, teachers can help students explore their attitudes and assumptions about communication—and global communication in particular. These discussions can be facilitated through the students' own workplace experiences and analyses of documents, or from reading about global workplaces such as VisionCorps. Additional activities might problematize the meanings of terms such as international, intercultural, globalized, internationalized, and localized. If students then depict visually their interpretation of these terms, these visuals can become the basis for discussing the complex implications of communicating globally while avoiding ethnocentrism. Students can also benefit from creating a fictional global business and its guiding mission statement, in which every word and interpretive nuance establishes meaning for internal and external readers. In much the same way they are changed (or they tell me haunted) by their ability to analyze documents critically, students can become aware of their assumptions and theories, and those of others, for communicating globally.

This chapter has no conclusion of overriding implications to be asserted from my research of VisionCorps or from the discussions of theories of corporate change and interpretive ethnography; VisionCorps employees and many theorists continue in their efforts to improve their approaches to global professional communication. I can, however, reiterate the issue that motivates this chapter and my teaching.[12] Metaphors that emphasize national borders or corporate boundaries are no longer helpful; too often, they reinforce cultural stereotypes and result in ethnocentric approaches to communication. Rather than concepts of unchanging, homogeneous cultures of fixed boundaries, communicators need to consider the indeterminacy of cultures and cultural membership. Audience or user-analysis is more complicated than stereotypical hints, especially when others too are given credit for global awareness.

REFERENCES

1. N. R. Blyler and C. Thralls (eds.), *Professional Communication: The Social Perspective,* Sage, Newbury Park, 1993.
2. L. Faigley, Nonacademic Writing: The Social Perspective, in *Writing in Nonacademic Settings,* L. Odell and D. Goswami (eds.), The Guilford Press, New York, pp. 231-248, 1985.
3. T. Peters, *Liberation Management: Necessary Disorganization for the Nanosecond Nineties,* Alfred A. Knopf, New York, 1992.
4. A. Toffler, *Powershift: Knowledge, Wealth, and Violence at the End of the 21st Century,* Bantam, New York, 1990.
5. C. Savage, *Fifth Generation Management: Integrating Enterprises through Human Networking,* Digital Press, Bedford, Mississippi, 1990.
6. E. Schein, Reassessing the 'Divine Rights' of Managers, *Sloan Management Review,* pp. 63-68, Winter 1989.
7. W. Wriston, *The Twilight of Sovereignty: How the Information Revolution is Transforming Our World,* Charles Scribner's Sons, New York, 1992.
8. K. Ohmae, Managing in a Borderless World, *Harvard Business Review,* pp. 152-161, May-June 1989.
9. S. Doheny-Farina, Writing in an Emerging Organization: An Ethnographic Study, *Written Communication, 3*:2, pp. 158-185, 1986.
10. R. L. Brown, Jr. and C. G. Herndl, An Ethnographic Study of Corporate Writing: Job Status as Reflected in Written Text, in *Functional Approaches to Writing: Research Perspectives,* B. Couture (ed.), Abler, Norwood, New Jersey, pp. 11-27, 1986.
11. S. Doheny-Farina, *Rhetoric, Innovation, Technology: Case Studies of Technical Communication in Technology Transfers,* The MIT Press, Cambridge, Massachusetts, 1992.

[12]My thanks to the students in my graduate seminars and undergraduate course in Writing for International Trade for researching and theorizing with me; Carol Leininger and Rue Yuan for the on-going discussion; Rich Freed for his thought-provoking comments; and Carl Lovitt for his suggestions.

12. C. Hansen, Writing the Project Team: Authority and Intertextuality in a Corporate Setting, *The Journal of Business Communication, 32*:2, pp. 103-122, 1995.
13. K. O. Locker, What Makes a Collaborative Writing Team Successful? A Case Study of Lawyers and Social Workers in a State Agency, in *New Visions of Collaborative Writing,* Janis Forman (ed.), Boynton/Cook, Portsmouth, New Hampshire, pp. 37-62, 1992.
14. G. Cross, *Collaboration and Conflict: Contextual Exploration of Group Writing and Positive Emphasis,* Hampton Press, Cresskill, New Jersey, 1994.
15. R. Freed, Postmodern Practice: Prospectives and Prospects, in *The Social Perspective in Professional Communication,* N. Blyler and C. Thralls (eds.), Sage, Newbury Park, pp. 196-214, 1993.
16. G. Gatien, Managing in the New Corporate Environment, *Technical Communication, 37*:4, pp. 415-419, 1990.
17. E. Rogers and M. Allbritton, Interactive Communication Technologies in Business Organizations, *The Journal of Business Communication, 32*:2, pp. 177-195, 1995.
18. E. Sopensky and L. Modrey, Survival Skills for Communicators within Organizations, *Journal of Business and Technical Communication, 9*:1, pp. 103-115, 1995.
19. E. Tebeaux, The High-Tech Workplace: Implications for Technical Communication Instruction, in *Technical Writing: Theory and Practice,* B. Fearing and W. Keats Sparrow (eds.), The Modern Language Association of America, New York, pp. 136-144, 1989.
20. C. Boiarsky, The Relationship Between Cultural and Rhetorical Conventions: Engaging in International Communication, *Technical Communication Quarterly, 4*:3, pp. 245-259, 1995.
21. M. Jacobi, A Cooperative Rhetoric for Professional Communication, in *Studies in Technical Communication: Selected Papers from 1993 CCCC and NCTE Meetings,* B. Sims (ed.), University of North Texas, Denton, pp. 145-162, 1992.
22. S. Katz, The Ethics of Expediency: Classical Rhetoric, Technology, and the Holocaust, *College English, 54*:3, pp. 255-275, 1992.
23. C. Miller, What's Practical about Technical Writing? in *Technical Writing: Theory and Practice,* B. Fearing and W. Keats Sparrow (eds.), The Modern Language Association of America, New York, 1989.
24. J. Perkins, Reconsidering Technology Transfer: An Antifoundational Perspective, in *Studies in Technical Communication: Selected Papers from 1993 CCCC and NCTE Meetings,* B. Sims (ed.), University of North Texas, Denton, 1993.
25. T. Weiss, Translation in a Borderless World, *Technical Communication Quarterly, 4*:4, pp. 407-423, 1995.
26. P. Drucker, The Coming of the New Organizations, *Harvard Business Review, 66*:1, pp. 45-53, 1988.
27. P. Drucker, *Post-Capitalist Society,* HarperCollins, New York, 1993.
28. F. Dubinskas, Cultural Constructions: The Many Faces of Time, in *Making Time: Ethnographies of High-Technology Organizations,* F. Dubinskas (ed.), Temple University Press, Philadelphia, 1988.
29. R. Rosaldo, *Culture and Truth: The Remaking of Social Analysis,* Beacon, Boston, 1989.

30. R. Rosaldo, Where Objectivity Lies: The Rhetoric of Anthropology, in *The Rhetoric of the Human Sciences: Language and Argument in Scholarship and Public Affairs,* J. Nelson, A. Megill, and D. McCloskey (eds.), The University of Wisconsin Press, Madison, pp. 87-110, 1987.
31. R. Moss Kanter, *World Class: Thriving Locally in the Global Economy,* Simon and Schuster, New York, 1995.
32. S. Martin O'Donnell, *Programming for the World,* PTR Prentice Hall, Englewood Cliffs, New Jersey, 1994.

CHAPTER 2

Information Systems and Technology in International Professional Communication*

DEBORAH C. ANDREWS

Information systems and technology make the global economy run. They are the essential connective tissue of international organizations, automating many routine transactions in finance, service, and training and bringing people together across borders of space, time, and culture. This chapter first briefly describes such systems, usually referred to as *IS* or *IT*. It then reviews how the technology causes organizations to reshape their structure and management and radically alters the nature of work. Finally, the chapter discusses the implications of these systems and changes for the teaching of professional communication.

INFORMATION SYSTEMS AND TECHNOLOGY

IS and IT are, broadly, computer-based methods for developing and delivering information. The applications arrive at the customer's computer on disks, CD-ROMs or, increasingly, directly online, from a growing number of companies

*Several of the ideas in this chapter derive from a presentation Heather Fox, of the DuPont Company, and I gave at the 1994 annual meeting of the Council for Programs in Technical and Scientific Communication. Heather also provided valuable suggestions on an earlier draft. In addition, the chapter draws on information I include in a textbook, *Technical Communication in the Global Community*, published by Prentice Hall. I am grateful to the College of Arts and Science at the University of Delaware for funding the preparation of case studies in international technical communication; that work, too, helped inform this chapter.

who specialize in these programs. Information technology is expanding rapidly in power, versatility, ease of use, and pervasiveness [1-5].

The simplest and most familiar form is probably voice mail. The evolution of voice mail nicely demonstrates how the technology grows up. A basic telephone answering machine transmits a recorded message to a caller and records a caller's message. A computer-driven system automates the delivery of information and allows interactivity. Voice mail can handle routine inquiries and responses, schedule students for classes, send off repair people to fix gas lines, report the status of frequent flyer accounts, place orders for the purchase of stock—and let you know if the order was executed. It can also embed details for novice callers and bypass routes for experts.

Other forms of IT, called "expert systems" or "artificial intelligence," automate information-gathering tasks. You feed information into the program in response to structured questions, and the program determines an answer. Relatively simple programs help you calculate your taxes, determine your career potential, or write a will. Bankers use more complex programs to automate the evaluation of customer loans. Physicians use medical expert programs to focus in on a diagnosis.

IT not only automates the delivery of information, it also connects people and information through an expanding number of networks. The major network, of course, is the Internet, a supernetwork of networks that speeds information and money around the globe. A rapidly developing part of the Internet is the graphics-based World Wide Web, home to information posted by government agencies, research institutes, and commercial vendors. People talk about the Web as if it were a place, with each information provider maintaining a "site" and various kinds of software allowing users to "visit" these sites. The expansion of the Web is so great that most estimates of both the number of sites and the number of users are highly inaccurate. Such networks create communities of interest across organizational and national borders.

While the Internet is an international, public delivery system, a new approach is the development of "intranets," sophisticated company-owned networks. Originally the means to link employees and company databases, they are now extending to include suppliers and customers. Eventually, intranets will link all three groups—employees, suppliers, and customers—through the Internet in an automated network. The network will allow companies to coordinate production and management at facilities located in many different countries. Developing intranets is one of the steps companies take as they "reinvent" or "reengineer" themselves. They introduce business process software to compile automated reports on such recurring information as inventory, sales, financial status of projects, payroll dispersement, and worker hours. The software can also integrate routine administrative tasks; it can scan résumés for the personnel office, cut checks for the accounting office, and add and delete people on a payroll.

Because of IT, research, sales, finance—any organizational or corporate transaction—can take place as easily around the world as around the corner. It creates an economy that is always open. A voice mail system, for example, accepts calls from any time zone and provides responses in several languages. In addition, information that moves as electrons rather than paper, as bits rather than atoms, moves internationally as easily as locally, often without the user even being aware of its origins in another country, and crosses borders without the fuss of custom's declarations or national postage. The London office of CitiBank, for example, processes all its transactions in South Dakota.

Such international distribution of operations will become even more appealing as the speed of the transactions increases. A new, faster Internet II is being developed by research institutions; three separate groups of researchers succeeded in transmitting information at a rate of one trillion bits a second through an optical fiber, the equivalent of 300 years' worth of daily newspapers in one second [6]. These high capacity links also help reduce the cost of communications, particularly in the United States, the site of stiff competition among providers. That competition results, in part, from a convergence of telephone, television, computer, and cable technology in what experts hope will be a "seamless" web of delivery for information and entertainment. A digital delivery system, too, means that words and pictures easily interpenetrate as pixels (picture elements) on a screen. High speed systems thus accommodate multimedia; words, images, sounds, animation, and numbers travel together. The following section describes some dimensions of the way that information systems have transformed the workplace and changed the nature of work.

THE TRANSFORMED ENVIRONMENT FOR WORK

In a clever sequence of phrases, an IT company describes its approach as *create, integrate, communicate.* The terms neatly highlight the need for connections among systems and people and for effective communication to make those connections hold. The picture of an integrated network, too, replaces the picture of a hierarchy as the image of the transformed workplace. Let's look at some implications of this new picture.

Changing Roles for Workers

First, IT changes the roles workers play. For example, computers and telecommunications are replacing many clerical and service jobs. Voice-mail and online databases answer customer questions automatically and interactively. IT replaces other people in the middle—insurance salespeople, travel agents, bankers, for example—as customers deal directly with suppliers, and it reduces levels of management by automating the information brokering and monitoring previously performed by managers-in-the-middle. It is in large part responsible for the rampant downsizing of the U.S. and, to a lesser extent, European companies in the 1990s. Moreover, IT allows companies to be more flexible in their locations, with fewer people covering wider territories. The network links them electronically and supports far more transactions per person than in a conventional organization. Finally, researchers have called the new worker roles "the end of the job." The technology encourages (some people say *forces*) workers to adapt to shifting roles instead of fixed positions. People increasingly work on project teams, sometimes as a company employee but often as an independent agent. The teams form and dissolve according to broad organizational goals and needs.

Effect on Productivity

Second, IT affects productivity, although the *direction* of the effect is much debated. Those who say it reduces white-collar productivity cite a "productivity paradox." Because the technology is at times (some people say "often") imperfect and overly complicated, companies may actually need *more* people simply to make the technology work. As an example of misapplied IT, one authority cites the attention several U.S. companies paid to ever more sophisticated inventory management software—while the Japanese developed a system of just-in-time manufacture that eliminated the need for such elaborate inventory management [7]. In addition to misapplication, another cause of a decline in productivity is *displacement of efforts.* "Many new things are being done by a work force of the same or larger size" rather than fewer people doing the old work for less money [8]. Empirical research shows, for example, that written communication, like e-mail, is more detailed and can take longer to produce than oral communication. When it tends to dominate over spoken communication in organizations, as it must when organizations are spread geographically, productivity slows.

IT also increases the channels of communication and can thus add to message overload, especially since messages arrive more frequently and, some people feel, more insistently. And being able to *produce* more information has also made people *want* more information, and thus information systems escalate the demand for data. That demand leads to a "management by numbers" mentality. One problem with such thinking is that the computer models may be simply wrong or misguided, yet they provide the "illusion of control" and accuracy [8]. A second problem is that such thinking then requires professionals to spend more time generating numbers to support their requests or idea and writing up those numbers in lengthy memos or reports. The technology simply expands the work. Ironically, while the technology has led to a reduction in clerical workers, some government data suggests that a displacement effect occurs in employment, too: "the hiring of more professional and managerial workers, at far higher rates of pay, who produce their own reports and correspondence on personal computers and answer their own phones" [8].

Researchers who see IT as contributing to productivity question the statistics. The apparent decline, they say, simply reflects a problem in measuring white-collar output. They also readily concede that the new technology destroys jobs, but argue that it creates more jobs than it destroys, at least in the long run. Over time, general rates of employment tend to remain about the same, while the relative demand for different kinds of workers shifts. In addition, according to a recent study by researchers at MIT, the average return on investment in company-wide information systems in the United States is 81 percent, compared with an average of 6.3 percent for everything else [9, p. 63]. Networked computers produce strong gains in companies that reshape how they do business and put aside old management habits.

Political and Ethical Implications

In transforming the environment for work, IT has broad political and ethical implications. Most important, information systems further reinforce the division of the world into the haves and the have nots. The gap continues to widen, for example, between highly paid, highly educated workers who run the technology and many low-skilled, low-paying, less secure, and part-time positions. Educators will perform an important role in narrowing this gap on both the broad social level and in their classrooms. In a brief, well focused list of recommendations for students, Robert Reich, the former U.S. Labor Secretary, itemizes ways in which education can contribute to narrowing that gap [10]:

1. Whether you work in an office or manage a crew that cleans it, you must be computer literate.
2. Keep your skills sharp and continue your education.
3. Ditch the ladder; catch the Web. Smart workers move along Webs, earning more from expertise, not from seniority.
4. To hone your skills, "network" with others in your profession.
5. Tomorrow's workers will function as "teammates." Learn to play all positions and win as a team.

The rest of this chapter expands on the implications of this advice by discussing strategies for teaching professional communication in the context of information systems and the global economy.

ACCOMMODATING INFORMATION TECHNOLOGY

Information systems are both a method of communication and a topic for communication. Professionals, regardless of their specialty, will use such systems. Professional communicators will both use and write about the technology. In addition, because of its prevalence, the technology often dominates discussions about any kind of organizational goal. Thus, as researcher David Dobrin notes, demonstrating how to accommodate such technology to its users will form the center of many, maybe most, communication tasks in the twenty-first century. A designer of information systems defines communication as "remembering what it was like not to know" [11]. Professionals who deal with such systems have to cultivate that memory. They have to provide the right information, through the right delivery system, to the right audience, at the right time, and at the right cost. A tall order. Attaining this goal requires some of the following skills that teachers need to help students develop:

- confronting ethnocentrism,
- analyzing audiences across borders of technology as well as culture,
- understanding the new genres of electronic communication,
- using technology to enhance independent work, and
- using technology to enhance international collaboration.

Confronting Ethnocentrism

At a recent Asia-Europe summit meeting in Bangkok, Malaysia's prime minister noted, "Asian values are universal values. European values are European values" [12]. That epigram puts a nice twist on a common problem among Americans: thinking their values are universal. In a wired world, such thinking is bound to cause trouble.

But such thinking comes easily, reinforced by the prevalence of English in international communication. Already the language of air traffic control, most scientific publications, and many business transactions, English, particularly American English, is becoming the language of the Internet. According to the director of a Russian Internet provider, "It is far easier for a Russian language speaker with a computer to download the works of Dostoyevsky translated into English to read than it is for him to get the original in his own language" [13, p. 5E]. There are probably now more people who speak English as a second language than those who speak it as their first. But the American bias does raise ethical issues. According to the Russian director, the American structure of the Internet is an "act of intellectual colonialism" that reinforces American cultural power. Students need to be made aware that the delivery system itself conveys a message.

In addition, they will need to avoid thinking that electronic systems are always the right and best technique for communication. In many locations, computers are still rare, electrical supplies are spotty, and phone lines are unreliable. That doesn't mean that the intelligence or business acumen of the people in those locations is any less than in the United States, but Americans often jump to that conclusion. Using IT well requires that professionals also assess when it's appropriate not to use such systems.

Analyzing Audiences Across Borders

Assessing when to use information systems requires an ability to analyze one's audience, a second vital skill for professional communicators. In the context of international information systems, such analysis incorporates issues of both culture and technology.

Students will need to see how their audiences are both like them—and different from them. The situation is sometimes ironic. Canadian management authority Nancy Adler notes the irony: "Organizations worldwide are growing more similar, while the behavior of people within organizations is maintaining its cultural uniqueness. So organizations in Canada and Germany may look the same from the outside, but Canadians and Germans behave differently within them" [1, p. 46]. Similarly, Dwight Stevenson notes, "Our technology may make us citizens of the world, but our rhetoric, our laws, our languages make us citizens of our own cultures" [14, p. 327]. A resurgence of nationalistic interests and patriotic organizations has accompanied the growth of the global economy, one of many sure signs that traditional differences among people remain [15].

On the other hand, David Victor cites a Tokyo management consultant's focus on the growing similarities, at least among the technological "haves." The consultant calls one group of such "haves" the "Triadians," residents of a triangle represented by Japan, the United States, and the European Union, whose "academic backgrounds, income levels . . ., life-style, use of leisure time, and aspirations are similar" [16, p. 9]. The Triadians tend to be young, and their homogeneity as sharers of information technology and as consumers increasingly means that they resemble each other more than any one national group of them resembles the elderly of their own country. This international elite can assume they have many interests and reference points in common, and thus they converse easily among themselves. A major integrated resource management package, SAP, for example, comes from Germany but runs operations in companies worldwide, including many of the biggest companies in America. A new rival is Dutch-based. Information workers are hardly aware of the national origins of such packages.

But this technologically sophisticated elite represents a thin crust of people worldwide, and thus assuming that all audiences fit that mold can be dangerous. One problem, as we've seen, accompanies an "English only" ethnocentrism. Individuals who rely on English as an international language may find that they are understandable. But they may not be *persuasive* unless they speak their client's language. Another problem derives from differences in age between those who use and teach the new technology and those who must learn. The bearers of the new technology are often the young; when they introduce that technology into organizations, they turn traditional mentoring patterns on their head. Their "know how," in Norbert Wiener's apt phrase, is often not matched by their "know what." So older workers, if their jobs remain after downsizing, may further resent the young. In addition, in cultures less youth-oriented than America and more oriented to "know what" over "know how," the role of young international consultants, for example, can be grating. One thirty-year-old American at a U.S.-based financial services company engendered much ill will when she restructured the back office operations in London by announcing that they needed to make "grown up" decisions. Students have to cultivate an ability to listen and to be polite as well as correct.

Finally, electronic delivery may simply not work with some audiences. As a vice president of Mercedes Benz noted, he is eager to deliver company service manuals to garages electronically, but he questions whether the garages have the technology to download them and whether mechanics have the skills or the inclination to understand them in that form. Messages have to reflect the technological sophistication of the audience.

In learning to communicate across cultures and technology, students should be encouraged to take audience analysis seriously and to create detailed, written audience profiles. Figure 1 shows an example of such a profile, composed by a team preparing information products for a computerized accounting system at the DuPont Company.

In addition to determining who the users are, developers of technology products need to determine what they want to achieve through the process. What problem is the product or system designed to solve for the audience? How will the audience measure success? Here is the goal statement for the DuPont accounting system:

> The information products will be used to train new GTDB users and to support them when they are on their own. User support is the most important because the users:
>
> - are already under time constraints and dealing with a lot of variables (time zones, languages, currencies, etc.). They need fast solutions to their usability problems to meet their deadlines.
> - depend on one another for quality financial information from the database. Bad data costs time at the very least and can potentially cause significant misstatements of financial information.

Understanding the New Genres of Electronic Communication

In situations like that at DuPont, where electronic delivery *is* appropriate, the technology can perform a powerful role in accommodating the needs of a diverse audience. To make that accommodation, professionals must draw on a third skill: an ability to use the new genres of electronic communication. IT and IS can deliver, of course, traditional document forms, like memos, letters, proposals, reports, and instructions. But its real strength lies elsewhere. As a delivery system, the technology fosters new genres as well as new approaches. In its interactivity

Primary Audience

The primary audience for the GTDB information products are the financial professionals (about 90 people) who currently use the COGS Transfer DataBase. These people are responsible for managing and communicating financial information for individual DuPont subsidiaries around the world. They have individual accounting responsibilities to their own companies and governments, but they also work together as part of the DuPont global accounting network to collectively report the financial position of the DuPont businesses and consolidated DuPont Company each month.

These users work in a high-pressure environment which is imposed by a monthly series of reporting deadlines. In dealing with their accounting responsibilities they must also deal with multiple time zones, languages, cultural factors (holiday, religions, attitudes), levels of technology, types of accounting systems and equipment, and currencies with fluctuating exchange rates. Their work is very interdependent. If one subsidiary is late in providing data, it could hold up other subsidiaries and the entire corporation. If users have difficulty using a system it will slow them down and cause greater anxiety for meeting the established deadlines.

A large percentage of this audience is not in the U.S. Although they all have some understanding of the English language, English is not the native language for many. We have mailed out a survey which is aimed at learning more about their cultural, educational, and language backgrounds, as well as other information. We plan to conduct some interviews to get more information. In addition, the Business Information group has a file of help line requests from the COGS Transfer DataBase.

The people who support the GTDB Help line are another primary audience. They work under the same time constraints as the GTDB users, acting as a problem-solving resource for them. There are currently two help line people, both located in Wilmington.

Secondary Audience

A secondary audience will comprise the following:

- Managers who need to know how the accounting systems work
- Auditors who need to confirm system procedures
- Business analysts

Figure 1. Audience Profile (GTDB is the new program; COGS is the program being replaced). Source: Susan Gallagher. © 1996 E.I. duPont de Nemours and Company. Used by permission.

and multimedia capabilities, it can answer an audience's specific questions, on the spot, in a style that suits that audience's preferences.

One new genre is the e-mail message, which is part letter, part memo, part phone call. Calling e-mail a "genre" begs the question, somewhat, about what characteristics define a genre [17]. But students do need to learn how to use e-mail well—efficiently and, to evoke the term again, politely. Another new genre is the World Wide Web site. Creating Web sites will become an increasingly common task for professional communicators, and even a new career. The technology fosters "mass customization," that is, the delivery of information so that readers can make one-of-a-kind, custom documents. In addition, pictures and videos can help to demonstrate what may be hard to describe in words, and sites can accommodate multiple languages.

Graphics are the core of most Web sites, so creating a site requires professionals to think in images. An image also underlies what might be called another genre of electronic delivery: the very smart map. A map format provides an excellent structure for displaying large amounts of digital information, either online or on a CD-ROM. A major technology for developing such maps is a Global Information System (GIS). Because of the role GIS increasingly plays in corporate decision making, it's worth a brief look at the technology. The market for GIS was worth about $75 million in 1995 and is rapidly growing. In emphasizing the importance of mapping, one U.K. authority notes that "75 per cent of all information in circulation has location as a common ingredient" [18, p. 11]. With GIS, a resource manager can decide about placing limits on development in a coastal area by overlaying several digitized maps: land cover in each of the last twenty-five years, municipal maps for the same period, and a map that represents statistics on shell fish. In the United Kingdom, GIS technology is popular, especially since highly accurate digitized maps are already available from the government's Ordnance Survey. North East Water, the largest water-only utility in Britain, uses GIS to monitor its 8,000 km piping network and let customers know about any emergencies, planned maintenance, or other special conditions. Marks and Spencer, a major retailer, uses GIS routinely in support of decisions about locating new stores, defining fuel-efficient delivery systems, and profiling customers and customer trends. Notes one Marks and Spencer researcher, "A map backdrop certainly helps us present complex scenarios in an easily-digestible form" to aid decision making [18, p. 11]. More important, it helps with detailed financial analysis through links with other software. A region in northern Spain is using GIS technology to deliver information to tourists and potential investors over the World Wide Web.

The map as metaphor provides an organizing principal for other forms of graphic display. One such form is a dynamic three-dimensional map. The user "flies through" information presented on different planes, at different scales. The program can assemble statistics, for example, on mutual funds in an information "space" in which different grids organize different kinds of information. Clear planes slice orthogonally through the grids to display several categories of information (like annual return on investment or sector weightings) or one category of information for all the funds. Once an audience adjusts to the conventions of such presentations, the program allows multiple access points to the data. The display demonstrates relationships dynamically and provides the user with interactive control, often independent of natural languages [19].

In this "post information" age, then, emphasis shifts from producing information to displaying it. Professionals in all specialties—marketing, engineering, science, manufacturing—will become increasingly responsible themselves, without much editorial assistance, for designing the communications that will deliver their information to colleagues, customers, and clients. Current concerns for layout and design of a page will give way to greater interest in indexing, navigational aids, and links. A big goal is developing technology to tame the chaos of the Internet. Researchers are working to build "autonomous agents" that will roam the electronic highway, visiting Web sites and other archives to do the user's bidding. The software becomes a kind of trusted servant. Currently available search engines are a step in that direction, but they are less comprehensive as well as less focused (thus they may proclaim, on the first screen of a search, "3000 matches"!).

Using Technology to be Independent

All these information displays can be delivered in an increasingly integrated network to an individual worker, at an individual workstation. And that has helped spark a decentralization of work. Decentralization requires professionals to exercise another skill: an ability to work independently. Many professionals in the twenty-first century will define themselves as "consultants" or, as their otherwise rural neighbors in the U.S. West sometimes call them, "modem cowboys." Like actors, they will work for a period in one show, then look for another, with periods of rest in between [7]. Some will become "mobile managers" in an organization's geographically dispersed workplace, keeping close to both suppliers and customers, on the supplier's or customer's home turf. Because even small companies have customers around the world, professionals may find themselves living extensively abroad and working, either abroad or at home, in a "virtual office." They will have more direct responsibility for decision making and fewer reporting lines. Computer-mediated conversations will frequently substitute for face-to-face communication.

Telecommuting (or "teleworking," to use the British term), and the more general concept of the virtual office, present attractive features, especially in such business sectors as finance and sales that draw significant profits from the international marketplace. But the technical systems that connect people easily across the globe are only part of a good working environment. Notes researcher Wanda J. Orlikowski, while "technology can enable ways of working, that doesn't mean work is now only technological. Work is still very much social" [17, p. 10]. Workers who rely only on mediated communication may miss the nuances, creative insights, and more risky speculation that can develop in face-to-face and less instrumental conversations. They are also less visible when it's time for advancement, a common complaint of expatriate professionals. And they may suffer from the incentive that computers and a global economy provide for a twenty-four-hour workday. A person who has customers or clients in every time zone may feel a not-so-subtle pressure to keep similar hours. Even from the company's perspective, supervising people over long distances taxes current corporate structures, as the example of rogue futures trader Nick Leeson in Singapore demonstrates.

In addition to a lack of visibility and an increase in stress, virtual work also raises issues of privacy. In a 1993 survey of employers conducted by a computer magazine, 22 percent of the respondents said they monitored employee voice mail, e-mail, or computer files—many without the workers' knowledge or consent [21]. Many employees worry about such corporate eavesdropping. The privacy of voice mail is a more open question than that of e-mail, because recent court cases have established the right of employers to read e-mail circulated on systems they own and operate. While most companies do not monitor e-mail, the fear of such oversight causes many professionals to avoid using e-mail to send the personal messages that might help them overcome the loneliness of long distance working.

Working virtually, then, requires that professionals reap the advantages of information technology without succumbing to its potential for isolation, invisibility, stress, and watchfulness. Rewards will come to those who are quick, flexible, and self-directed. To prepare for increased independence in their careers, students need to:

- learn techniques for finding and filtering information through traditional as well as electronic searches,
- sample current technologies to *learn how to learn* the very different technologies they'll probably encounter on the job,
- practice consulting skills, that is, sizing up situations rapidly to determine where the strengths and problems lie. Case studies will help teach these skills, especially cases that present the complexities and ambiguities of international deliberations,
- develop a taste for uncertainty and rapid change, and
- become comfortable listening and talking at a screen, presenting themselves effectively in that mediated environment.

Using Technology to Collaborate

In the transformed workplace of networked organizations, students will also have to balance their skills at using technology to be independent with a seemingly opposite skill: using technology to collaborate. Professionals will increasingly move in and out of teams as projects and goals demand. Those teams, too, draw together people who do not necessarily share the same values, visions, language, and expectations about how things run and what documents should look like. Here are two current examples of international team approaches:

- The U.S. National Science Foundation sponsors an electronic research effort, called a "collaboratory," through which some 30 space scientists from California, Maryland, Michigan, and Denmark conduct studies based on data gathered remotely from five instruments in Greenland. The system feeds data and commentary to each researcher's terminal. Each screen has a window that displays the data in various visual forms as well as message boxes for sending and receiving text. A small menu allows researchers to shift to different instruments for gaining data. Such collaborative technology has obvious benefits, including providing an abundance of observations without the need for costly trips to Greenland. In addition, "More eyes on the data increases the chances of a 'Eureka!' event," according to one scientist [22, p. A16].
- To market a product in Japan, a Chicago-based pharmaceutical company needs to comply with multiple Japanese regulations and protocols in business and medicine. Because of physiological differences between Americans and the Japanese and differences in medical practice, the Japanese government requires additional testing of drugs beyond those necessary in the U.S. For these reasons, the Chicago-based company has formed a joint venture with a Japanese company. Members from each partner collaborate to design an appropriate product and an appropriate marketing approach.

Advantages of Teamwork

Each of these examples demonstrates how a team approach helps achieve a specific goal. In the first example, experts in the same specialty (space scientists) divide the work to speed up the research process and share a rich database that would be too costly for each to develop. In the second example, a joint venture helped a Chicago company and a Japanese company create a new business neither could have achieved on its own. Local experts helped those from away adapt to the new market. These brief descriptions point out two major features of collaboration: the pooling of specialized knowledge and the strong role of information technology as both topic and tool for teamwork.

Faced with the task of solving a complex problem, a team can assemble specialists in a variety of areas to focus on the issues at hand. Such expertise is particularly valuable when projects cross national borders and thus require familiarity with local conditions and the local language. For example, because legal systems differ across the world, multinational organizations often need to work with local experts to prepare such documents as contracts and operating manuals that use symbols and include warnings and other cautionary statements to meet regional government standards and codes and conform to varied liability laws. Companies seeking to expand their information systems also often "outsource" the development of those systems to expert vendors who then become part of an otherwise internal company team.

Technology for Teamwork

While joint ventures and other collaborations often begin with a face-to-face meeting, electronic technology provides a less expensive and more flexible medium for continued conversation. Simple technology for collaboration includes phone calls, faxes, and e-mail. More sophisticated applications let geographically dispersed team members share documents, commentary, and databases. Team members contribute to the project discussion at different times that match working hours in their own time zones; the discussion then builds as each new member logs in. Or team members may work on a document simultaneously, adding their comments in windows or boxes separate from the text. In either form, the software stores drafts and comments to establish an explicit group memory.

Teleconferencing systems that link people at remote locations through telephone or satellite hookups also enhance the opportunity for groups to discuss issues. Full-motion video conferences can be beamed worldwide among specially equipped rooms so that participants can see and hear one another as well as share displays on computer screens. Increasingly, too, personal computers can be linked across several geographically remote locations to provide a conference-room-on-a-desk. Small video cameras photograph each participant for display in one window on the screen while data and comments appear in other windows.

Teamwork across Borders

Whether they communicate face-to-face or screen-to-screen, professionals must draw on special skills to communicate effectively, especially on multinational teams. Ethnocentrism may get in the way. So, too, may differences in the way people of different ages or a different gender expect to be treated in their own culture. Patterns of relationships also differ from culture to culture, so Koreans, for example, who value strong kinship bonds may expect similar tenacity and longevity in a collaborative project, whereas people from more individualist cultures can form and disband groups more readily. Folklore sometimes suggests those values, as in a Dutch expression, evoked by a Dutch professional when a joint authoring project with U.S. colleagues wavered: *samen uit, samen thuis.* It means "together out, together home," that is, you begin a project together, you finish it together.

Teamwork requires people to share information, but some people naturally resist such sharing. That refusal sometimes reflects a personal or cultural value in not wanting to confront others and speak up publicly. It may reflect the fear of making a mistake and the need to save face [23, p. 57]. Or it may derive from an unwillingness to disclose proprietary information in, for example, a discussion of a joint venture among otherwise competitive interests. If team members hold their cards too closely, the effort collapses.

To prepare for increased collaboration in their careers, students need to:

- practice teamwork in the classroom through competitive, task-driven projects,
- use e-mail and other available technology to support that team practice, and
- learn how patterns of collaboration and patterns of persuasion differ from culture to culture.

FUTURE WORK

Information systems and technology will continue to transform the way the world works [24, 25]. Today's students will play an important role in the transformation when they become professionals who create and use the systems. Today's teachers should help them develop the skills to play their role well. The technology may seem daunting, overwhelming, or even, from a certain point of view, beside the point. While teachers should be familiar with IT, and curious about it, they need not fear even if they remain life-long IT novices. What's critical is not mastery of the system, but an understanding of how systems and users interrelate. This chapter has pointed out key skills students must learn to make those interrelationships effective. Those skills will help them reap the benefits of technology in the global economy. But there are many traffic jams on the information highway, as well as accidents and a whole lot of debris. At the same time that the potential for global communication has increased exponentially, so has the potential for global *misunderstanding*. We need to know more about many issues concerning the interactions of people from different cultures through information technology, for example:

- how to make the technology more intuitive and easier to use,
- how, then, to design better interfaces between the user and the technology,
- how to preserve the social values of work in a virtual workplace,
- how to manage multicultural teams, and
- how to reduce the gap worldwide between the technologically sophisticated haves and the technology starved have-nots.

Developing information to solve such problems is not easy, but these are challenges our students will face. In our teaching and research, we need to make the technology more flexible, better adapted to users, and more humane.

REFERENCES

1. N. J. Adler, *International Dimensions of Organizational Behavior*, Kent, Boston, 1986.
2. J. D. Bolter, *Writing Space: The Computer, Hypertext, and the History of Writing*, Lawrence Erlbaum Associates, Hillsdale, New Jersey, 1991.
3. N. Negroponte, *Being Digital*, Knopf, New York, 1995.
4. S. Turkle, *Life on the Screen: Identity in the Age of the Internet*, Simon and Schuster, New York, 1995.
5. S. Zuboff, *In the Age of the Smart Machine*, Basic Books, New York, 1988.
6. A. Pollack, Speed Record: One Trillion Bits a Second, *The New York Times*, p. D2, March 1, 1996.
7. A. Sampson, *Company Man*, Random House, New York, 1995.
8. P. A. Attewell, The Productivity Paradox, *The Chronicle of Higher Education*, p. A56, March 15, 1996.
9. Wiring Corporate Japan: Doing it Differently, *The Economist*, pp. 62-64, April 19, 1997.
10. G. Burkins, Work Week: Reich's Rules, *The Wall Street Journal*, p. 1, June 4, 1996.
11. R. S. Wurman, *Information Anxiety*, Doubleday, New York, 1989.
12. Asia and Europe: Friends Apart, *The Economist*, p. 33, March 9, 1996.
13. M. Specter, World, Wide, Web: 3 English Words, *The New York Times*, pp. 1E, 5E, April 14, 1996.
14. D. W. Stevenson, Audience Analysis Across Cultures, *Journal of Technical Writing and Communication, 13*:4, pp. 319-330, 1983.
15. F. Trompenaars, *Riding the Waves of Culture: Understanding Diversity in Global Business*, Irwin, New York, 1994.
16. D. Victor, *International Business Communication*, HarperCollins, New York, 1992.
17. J. Yates and W. J. Orlikowski, Genres of Organizational Communication: A Structurational Approach to Studying Communications and Media, *Academy of Management Review, 17*:2, pp. 299-326, 1992.
18. P. Ireland, Top of the Shopping Lists, *Financial Times*, p. 11, October 4, 1995.
19. R. S. Wurman (ed.), *Information Architects*, Graphis, New York, 1996.
20. M. O. Kirk, The Virtual Office Bumps Into Some Very Real Limits, *The New York Times*, p. F10, March 3, 1996.
21. F. A. McMorris, Is Office Voice Mail Private? Don't Bet on It, *The Wall Street Journal*, p. B1, February 28, 1995.
22. B. T. Watkins, A Far-Flung Collaboration by Scientists, *The Chronicle of Higher Education*, pp. A15-A17, June 8, 1994.
23. D. S. Bosley, Cross-Cultural Collaboration: Whose Culture is it, Anyway? *Technical Communication Quarterly*, 2:1, pp. 51-62, 1993.
24. N. Nohria and R. Eccles (eds.), *Networks and Organizations: Structure, Form, and Action*, Harvard University Press, Boston, 1992.
25. M. O'Hara-Devereaux and R. Johansen, *GlobalWork: Bridging Distance, Culture, and Time*, Jossey-Bass, San Francisco, 1994.

CHAPTER 3

Using Technology to Support Global Drug-Development Teams

STEPHEN A. BERNHARDT

Drug products reach the market by successfully negotiating health authority gatekeepers (the Food and Drug Administration in the United States, other authorities in other countries) and by having well substantiated claims for new or more effective disease treatment. The New Drug Application (NDA) is the major document set that presents the case for approval to authorities. An NDA represents an immense documentation task, made difficult by multiple global markets with corresponding health authorities and regulations, a changing regulatory environment, and intense pressure to deliver drugs in a timely and efficient manner within tight budgets and timelines. Technology holds great promise for global teamwork; indeed, it is only through technology-supported teams that an articulate vision of global drug development can be realized.

In this chapter, my purpose is to discuss document development and associated technologies that support global teamwork, based on my work within the pharma industry. From May 1995 to June 1996, I was on leave from my professorship at New Mexico State University to work as a full-time consultant to the pharmaceutical industry. As Senior Consultant for Scientific Services, Franklin Covey, my task was to work with globally dispersed teams as they developed new drugs for market. I was based in Basel, Switzerland, working primarily with one large pharmaceutical company and its affiliates in France, England, Germany, Japan, and the United States.

The limitation here is that I will be writing from experience in one industry. But much of what I say, I believe, can be cautiously generalized to other research and development settings. Pharma development is more document intensive than many industries, and it is much more research intensive than many, with vast resources committed to generating a stream of new drug products, which is the only way a large company can be successful. I hope my audience, whom I envision as other students and professors of professional communication, will find value in a description of document development and its attendant technologies within an industry that has not been a subject of our literature.

Those of us who teach professional communication are typically attuned to application: how what we learn in industry settings can be applied to our own classes and curricula. So a second purpose here is both to confirm current practice and to suggest appropriate emphases in the preparation of students for work as either professional communicators or as professionals who communicate.

It is not meaningful to discuss technologies apart from their contexts of use, and so I will begin with discussion of document development in general and the specific nature of our consulting intervention, and then move to a discussion of global technologies that facilitate drug development within the pharma industry.

SETTING A CONTEXT: DOCUMENT DEVELOPMENT IN THE PHARMA INDUSTRY

The pharma development process delivers drugs from the research lab to the market. The process is notoriously long—some twelve years on average from discovery or recognition of a new drug substance to marketing a drug product. The research is risky and costly—many potential drugs are abandoned during research and development, and those that make it to market do so on multi-million dollar development and marketing budgets. The burden of paperwork associated with a project is enormous. A recent final NDA (New Drug Application) submitted to the FDA (Food and Drug Administration) was a dossier of some 600 volumes containing 120,000 pages, and some of the larger NDAs are over 200,000 pages.

Submission of an NDA is a significant milestone, but not the end of development. Presentations to health authorities follow, with both written and oral question/answer sessions; various publications make the research behind the product known to the scientific, medical, and insurance communities; and extensive marketing plans and product literature position the drug product and ensure that it reaches its customers. Our consulting intervention was intended to improve the quality and efficiency of communication surrounding drug development through coaching teams in document processes, encouraging good team communication, and making effective use of supporting technologies.

There are many threats to the smooth movement of a drug product from laboratory through development to market. Most notably, after years of investment in development, clinical trials may fail to provide compelling evidence of the drug's effectiveness, or fail to demonstrate that the clinical benefits outweigh the safety risks associated with the drug. With regard to the NDA itself, pharmaceutical writers must confront the challenge of producing an immense document, written by many individuals and aimed at readers with widely varying backgrounds and purposes. Currently, NDAs tend to be too much the product of dispersed and private energies. Too frequently, they represent the separate efforts of different research and development groups rather than an integrated dossier—the product of a team that shares a common vision, strategy, tactics, and language.

A rule of thumb says that for drugs that reach the market, the top 100 sellers will generate average eventual revenues of $1 million per day. The number of days a drug generates such peak sales depends on its patent protection, which, when it expires, allows generic competition. The company registers the compound out of the laboratory so its claim is protected throughout development, and the patent clock begins ticking as soon as the compound is registered. It is, therefore, immensely important to bring a drug to market quickly.

The successful filing of an NDA is threatened by inconsistent claims, poorly coordinated presentation, or confusing organization. If teams can write more quickly and dependably, if dossiers are more coherent and better coordinated, and, if evaluation by readers can be made more efficient and more likely to be positive, then the rewards for the pharma company are large.

DRUG DEVELOPMENT TEAM PROCESSES

The pharma industry has moved toward empowered teams that are assembled from across the organization to shepherd a drug through development to filing and beyond. Close collaboration across functional areas allows teams to develop dossiers that present a unified argument for a drug product, characterized by consistent messages, a focus on important issues, and a compelling presentation of the data and supporting rationales for approval. Close collaboration helps ensure smooth transitions from lab, to development, to market.

Franklin Covey has worked with numerous pharma companies and health authorities to identify and measure specific document quality standards for NDAs and to recommend certain ways of working to our clients. Based on the formal evaluation of multiple dossiers from various pharma companies, on long-term client interventions, and on research inside the FDA, Franklin Covey recognizes that high quality NDAs share these criteria at all document levels:

- The dossier is driven by a clear sense of purpose, organized at every level to emphasize strong, consistent messages about the drug. The dossier explicitly presents and argues a position.
- The dossier and its documents place conclusions up front in emphatic positions, with the presentation of data following deductively from prominent interpretive positions.
- The dossier focuses on critical issues, placing them in context and presenting clear responses to the issues. The dossier does not hide or bury issues in the

documents but turns a bright light on them and offers a considered response, with a stated rationale and support.

- The dossier works as a coherent whole, with consistent messages developed and linked across sections and plenty of signposts as to how the dossier is organized.
- The dossier makes good use of its visual dimension, presenting as much information as possible in visually interesting, compelling, and informative ways.
- The dossier is presented in effective and complementary print and electronic versions.

These are high standards and not at all typical of many NDAs. More typical is the dossier that is a mere compilation of reports and data, where the reader struggles to interpret the data and where conclusions are buried deep within documents, shrouded by a cascade of "objective" facts and a homogenous pool of detail. In the typical dossier, messages are missing or inconsistent across documents, readers are expected to develop their own interpretations, and there is no close harmony unifying the various elements. Frequently, the writer succumbs to the temptation to divert attention from troublesome issues of safety, problems with development, or inconsistencies in the data, hoping that if issues are downplayed they might escape attention in the immense documentation of the NDA.

Developing dossiers that meet high quality standards demands new forms of teamwork. It is increasingly typical across the industry to find cross-functional teams forming early in the development process, with representation from all crucial areas of development:

- **the research laboratory:** where new molecular entities are first discovered, isolated, or synthesized and judged to have potential for development,
- **preclinical research:** where new substances are tested for tolerability, safety, and effect, first in animals and then in healthy volunteers,
- **technical development:** where the compound synthesis is refined and scaled up from the grams and kilograms the lab has produced to large-scale manufacturing campaigns involving tons of active ingredients and various formulations of a drug product,
- **clinical research:** where controlled studies of pharmacology, effectiveness, and safety are conducted with increasing numbers of patients in a range of doses to establish eventually through two large, well controlled studies the effectiveness and safety of the drug product and thus justify approval,
- **regulatory:** where the drug product is ushered through the approval process with various health authorities on the basis of documentation, hearings, and site inspections, and

- **business and marketing:** where markets are forecast, pharmaco-economic models developed, launch plans prepared, and the product is taken quickly and effectively to market.

Cross-functional teams represent different discourse communities—people who belong to different, well established communities within the company. These groups represent differing educational backgrounds (pharmacology, statistics, chemistry, physiology, biology, marketing: all highly educated with many PhD or medical degrees represented), they speak different professional languages, and they have different epistemological positionings. Team members know, for instance, that statisticians speak a different language and possess intimate knowledge of highly specialized argumentative conventions, and team members appeal to the statisticians for permission to pursue certain lines of reasoning, to argue from data in warranted ways, and to formulate the strongest legitimate data-driven claims. Team members know, further, that preclinical scientists will only be willing to make cautious extrapolations from animal data to humans, even when it seems to other specialists that the comparisons are logical or compelling. Additionally, teams are likely to represent different cultures and languages. It is not unusual, for example, to find that the chemists are German, the clinical researchers French and American, the regulatory professionals British, and for the team to include Swiss, Japanese, or other nationalities. The team must be globally representative to address the critical issues related to global approval and marketing. So there are many languages and assumptions in play about how language works, what good documents look like, and how arguments should be made.

Which of these differences are most salient in how teams interact? Education and training? Job function or position? Nationality or language group? I cannot say with confidence on the basis of either data or hunch what most influences communication across the boundaries of various discourse communities. We talked frequently among our team of consultants, agreeing that in some situations, cultural membership seemed to be a strong influence, while in other situations, it was obviously a matter of, for example, chemist vs. clinician that seemed to pose the greatest obstacles to communication. We did agree that team members could be encouraged to talk about the boundaries, could learn to understand differences within the team, and could talk and work in ways that tended to get work accomplished efficiently. We recognized there was little need of invoking stereotypes of nationality or personality, and that the team could refuse easy categorizations of their fellow team members to confront, instead, the particular issues and conflicts that inevitably arose within the development process itself.

DOCUMENT DEVELOPMENT PROCESSES

Franklin Covey, in its contracts with various pharma companies, attempts to bring a message-oriented, issue-driven approach to drug documentation development. The process encourages drug-development teams to meet early in the development process to articulate a shared vision: the team members try to put *in* words *on* paper what it is they hope to be able to say about the drug product when it enters the market. We asked certain heuristic questions to facilitate this process.

- What *Messages* can the team deliver about the need for the drug, its efficacy, safety, indications, manufacturability, and marketability?
- What *Issues* are likely to confront the product, to hold up its development, cause safety problems, raise costs, delay delivery, or result in a product that is unable to gain a market share?
- What *Responses* does the team propose for addressing the issues?
- What *Support* does the team have or need to gather? How and where will the data, rationales, and arguments be marshaled in support of the application?

The idea is to let the documentation drive the science, so that the wide range of activity done throughout the development process arrives on time at shared milestones and delivers a product that meets a real need and captures a significant and predictable market share.

Within our practice, the work to articulate the issues with the team results in a table of team messages and issues (here represented in Figure 1 with faked data from two areas, the first preclinical and the second technical):

Messages	Issues	Responses	Support
Cell turnover does not translate into histopathological changes in animal or man; observed changes do not signal precancerous condition.	Study NS5713 found colonic proliferation of mucosal cells in the lower intestine.	Condition only seen in one study, where a high calcium diet probably stimulated change. No histopathological changes in 3 months.	Give history of studies developed in clin pharm and tox; FDA agreed to duration of study. Continue monitoring.
The relatively low melting point of the active substance does not cause a problem for shipping, storing, and marketing the drug product.	Active substance melts and agglomerates at 37-40°F.	Active needs temperature controls on shipping and storage. Use hamsters (temperature tags) to insure continuous temperature monitoring from synthesis through production.	Detail and assure cool room temperature handling from milling through synthesis.

Figure 1. Sample Message and Issues Table.

The table serves as a tool for visual thinking, guiding the team through development. The table serves as a focal point for the team as it is revisited regularly, and it provides a common ground for understanding that can bridge various discourse communities. The table fixes development issues, at least temporarily, in a single place where team members can view, debate, refine, and agree on what they hope to be able to say and how they hope to be able to respond to the difficult issues. The table grows with the project, as the team adds new issues or reconfigures the messages in response to the support, or lack of support, provided by incoming data and developments within the market.

From the table constructed by the team grow various documents during development. Scientists meet with regulatory and marketing team members to plan the drug product development process through prototyping. A *document prototype*© is an early map of a document, an elaborated outline that aligns messages and issues with various sections of a document.

The key document that is prototyped early on and that drives the development process is the target profile. This document prototype© eventually morphs into various forms of product information for prescribers and patients; for example, into the *package insert* in the United States. (The insert is that piece of paper packed into every drug product, meant to guide prescribing physicians and often read by those patients who wish to be well informed.) The package insert constitutes the contract between the drug company and the health authority, governing what can be claimed, what the indications (disease conditions) are for which the drug is usefully prescribed, how effective and safe the drug is, and what dose is appropriate. Every line of the package insert must have specific, referenced support in reported research. By developing proto-forms of this document early in development, our goal is to encourage teams to be document-driven in their science, with the market product reverse-engineered from the goal (what is written in the package insert) to research and development processes intended to justify and support those messages.

Document prototyping© forces scientists to think like business people; it forces team members to agree on where the project is headed and to develop strategies for delivering the key messages and dealing with the critical issues in direct ways. The process is very similar to what we see emerging in the computer industry, where the interface is prototyped first on the basis of what the customer tasks are, or where the documentation is written up front to guide the software development. In our consulting work, we talk about "right-to-left" thinking, because we start with the end of the process (the right side of the project timeline) and map out what must be there (the target profile). We then move backwards on the development timeline to place the milestones, deliverables, critical path activities, key studies, and so on, so that the final picture shows development moving left to right in ways engineered to arrive at a predetermined goal. The process is also similar to those we are familiar with for document production, working from the deadline back through production, review, drafting, and research to arrive at a timeline that can keep a writing project on schedule [1-3].

Throughout development, the cross-functional team meets frequently, taps the diverse perspectives and expertise of the members, jointly "owns" the developing dossier, and is responsible for seeing that it meets document quality standards. This cross-functional team approach contrasts with typical ways of working through functional areas and publication centers, where individual writers—clinical documentation specialists, for example—are handed the protocols and statistical reports and are expected to write up the various sections of the dossier on their own with occasional routing of the document for review comments and approval. In a team approach, the final dossier reflects the team's efforts and success, rather than being a compilation of individual documents gathered and bound for submission.

The document processes we implement attempt to encourage team ownership of documents. The high-level reports and summaries are prototyped by the team working as a group. Practically all documents are drafted by specialists working together, with the goal of creating documents that are built from the ground up to be coherent, linked to other documents, and mutually reinforcing. This practice contrasts with the traditional "Everybody write a section and then we will paste them all together" (see Couture and Rymer for distinctions among different forms of document collaboration [4]). Teams revisit prototypes several times during development to create more advanced drafts, to agree on the length of various sections, to develop the rationales and support, and to identify areas that need further scientific work. Area reviews (by specialists with similar backgrounds) lead to later team reviews, where cross-functional expertise ensures that issues are linked and treated in consistent and harmonized fashion. Final team reviews bring together those who, by virtue of their positions, must officially sign off as the final step in issuing authoritative reports in the name of the company.

Most drug products are developed for a world market, so dossiers must be simultaneously prepared for the European authorities (with varying requirements for different countries), for Japan, and for other countries around the world, with document delivery customized to the particular registration requirements of each country. Some countries have more stringent testing standards than others, some countries accept a drug if it is already approved in certain other markets, some countries expedite approval for certain socially important drugs (as for AIDS), some markets do not tolerate certain side effects, while other markets do not tolerate large pill sizes or certain routes of delivery. Development demands careful coordination of knowledge that is highly dispersed across a global company.

All during the long development process, the global team produces a wide array of written and face-to-face communication:

- Frequent e-mail and memos coordinate the teamwork.

- Internal plans, proposals, and reports keep the project budgeted and within the company's portfolio.
- Numerous written filings and oral briefings are required with health authorities.
- Research protocols, interim reports, published papers, conference presentations, and final reports guide the science.
- Huge volumes of data from manufacturing scale-up take the substance under tightly documented controls from lab synthesis to large volume production and packaging.
- Electronic versions of the information make the results of clinical trials of safety and effectiveness accessible for analysis by health authorities using their own statistical applications.

Overall, drug development represents a massive documentation project. The intervention methods described above attempt to tap the collaborative power of teams to deliver high quality documentation efficiently and with appropriate strategic thinking.

GROUP PROCESS NEEDS

In addition to document tasks, a cross-functional NDA development team has many social tasks which technology can support. It is important to realize that designing technology for teams means engineering an environment to support relationships and encourage collaboration.

Teams obviously need a space to meet frequently and work. They need to discuss and come to decisions about various logistical issues: who will do what by when with what resources under what procedures. The team needs to evolve a project timeline, manage a budget, assign tasks and deadlines, and track tasks to completion, paying particular attention to those tasks on the critical path. For a project as complicated as delivering an NDA, the best planning is subject to constant revision as data come in, production problems are discovered, questions arise with health authorities, or market factors develop. When the development project is global and involves planning for simultaneous submission in a world market, the need is heightened for good technological support for the formal processes of project planning and management.

In addition to these more procedural and formal group tasks, there are more social, less formal issues that groups must address. There must be consensus among team members about what their overall goals are, what their standards are, how they will work, and how decisions will be made. It is a complicated and delicate business to reach consensus about what kind of team they belong to and what their obligations are to the team, especially in organizations where staff have obligations to line management as well as team management. These social tasks of arriving at group definition and consensus on goals and ways of working should not be underestimated. The best teams achieve a cohesiveness, a bonding through working toward a shared goal, and they find energy and identity in belonging to a project team. They feel recognized and rewarded for their individual contributions to the team, and the team itself feels as though it is recognized and rewarded by the larger organization.

The processes of document development and teamwork described above are idealized, representing how the process unfolds with the teams and leaders most willing to change traditional practice. What I have described represents the goals of our consulting intervention, not necessarily the outcome for any particular team.

We recognize the success of interventions through such indicators as the following:

- We helped a major project transition from reliance on external consultants to an internal department function, with new hires, a regular budget line, and a place on the organization chart,
- We witnessed dramatic improvement in document quality, as measured against benchmark standards established at the beginning of the project, based on document quality scores for some 60,000 pages from the four different dossiers, and
- We watched employees use the language, theoretical constructs, and practices of the intervention in natural contexts of team interaction. When, as a consultant, you begin to hear people speaking your language, making your arguments, and urging fellow team members to follow practices encouraged by the project, you know you have achieved some measure of success.

The next section will examine ways in which technology is a key component of the project, a lever of change, and a natural way to encourage global, team-based, document-driven development.

CHOOSING ENABLING TECHNOLOGIES

Pharma companies are interested in new tools that can help global drug-development teams deliver quality documentation in timely fashion. The emergence of groupware technologies over the past few years holds much promise [5-8]. Technology can help bridge distances, allowing NDA teams to "virtually co-locate," to find ways to meet and work together without physically being in one place. Traveling is costly, inefficient, and results in considerable wear-and-tear on a team. Physically co-locating teams triggers serious hardships for employees and their families and costly expatriate compensation. So there are very pragmatic rationales for meeting and working electronically.

The definition of the technological systems to support global, cross-functional teams follows the need to support both collaborative document tasks and group communication processes [9]. Shrage usefully identifies and describes the *shared space* that collaborative technologies can create, wherein team members focus their intellectual energies around project goals [10]. The essential step for Shrage is the movement from assignment to a team to the true collaboration of individuals in the act of leveraging their intellectual energies to accomplish work. The title of his book, *No More Teams,* reflects his disparagement of many so-called teams which are actually people assigned to work on something together and who never realize the benefits of true collaboration (and the title of this same work as previously published, *Shared Minds,* reflects what Shrage considers the goal and benefit of true collaboration). In our writing classrooms, we might compare collaborations where individuals contribute sections or perform discrete functions (researching, reviewing) with situations where a team actually collaborates by inventing, drafting, and revising in close partnership.

A wide range of software and systems can coordinate and expedite the activities of globally dispersed teams seeking to form collaborative working relationships. Tools like voice-mail and e-mail have become so ubiquitous that we sometimes fail to remark that they are really groupware tools. Nevertheless, how teams use such tools has important consequences for the structure and success of teamwork [11, 12]. Spreadsheets, charting, and project management software can keep teams on track toward planned outcomes, and really should figure into a discussion of groupware. My interest here is a bit narrower, however. The following section highlights several key tools currently in use that help create shared space for team collaboration involving key communication activities of documentation creation and presentation.

Videoconferencing

Videoconferencing is the most widely sought and most immediately comfortable technology for bringing together globally dispersed teams. For the team, videoconferencing is a "turn it on" technology, as familiar as television sets and microphones. While videoconferencing is quite expensive (phonelines carry the signal and can run to $40,000/month for a multipoint connection), pharma companies are accustomed to providing videoconferencing studios and operators, and telecommunications companies are accustomed to providing support and troubleshooting.

Videoconferencing is critical to drug development, as it allows teams to work together on a daily basis, even though members are located in different countries or cities. For traditional meetings, where work is conducted by talking and keeping minutes, videoconferencing is widely accepted as an adequate substitute for face-to-face communication. Videoconferencing gives a sense of personal immediacy—others present, personable, and knowable—almost like a conversation. High bandwidth connections (for example, dedicated T1 lines working at >1.5 Mb/sec), although expensive to run, offer a sense of personal immediacy that greatly improves communication over the slower, jerkier video that some of us have seen on computers or older video systems.

Good sound is possible with high quality microphones and speakers. Sound quality should not be underestimated in its contribution to distance communication, especially when participants come from different language groups. More than one of our non-native English speakers observed that sound quality was more important to understanding than the video image. Even good video and sound transmission, however, are frequently marred by poor connections, timing lags, or feedback from the audio system—not to mention that the systems don't work well when piles of notebooks, reports, and notepads cover the microphones.

Precisely because it mimics face-to-face communication, videoconferencing is deceptive, often lulling team members into false feelings that good communication has taken place and that substantial work has been accomplished (somewhat like leaving a classroom thinking, "That was a good discussion," without wondering how students might be changed, or not, as writers or readers). In fact, video frequently impedes good communication and has not proven to be anything like a full substitute for face-to-face meetings [13]. Though participants seem immediately present, much non-verbal communication is lost—many gestures, facial expressions, and discourse structuring cues are likely to be missed or misinterpreted. If participants are to be heard, they must speak forcibly, and the attendant interpretation can be that the speaker is being aggressive or obstinate, which, in turn, encourages others to adopt strategies for dealing with boors. Those who speak in normal tones (and again, especially across language groups) risk not being heard. My estimate would be that in videoconferences involving two sites with five to eight team members at each site, some 25 percent of the utterances are not understandable at all to those at the other site and that only a few, perhaps 10 percent of the utterances, are heard by everyone. This means that conversational repairs are continuously being made, with requests for speaking more loudly and repeating what one has said occurring on every fourth or fifth utterance. Those who interrupt for repairs risk being perceived as offensive or nagging. For whatever reason, those who speak in small voices tend to do so even with constant requests from the team not to, and though they might begin their next utterance in a pronounced voice, their volume quickly tails off. It becomes aggravating to everyone never to be able to understand certain people on the team. Such problems are evident in face-to-face meetings across language groups, but they are exacerbated by the remove of videoconferencing.

Many other communicative actions are muddled by videoconferencing. The pattern of talking over another's sentences, or finishing another's utterance is devastating to the conversation, as are side conversations or *sotto voce* moves. Turn-taking is much more difficult, and controlling a conversation through pauses or other subtle signals does not work well. I frequently witnessed participants whose facial expressions indicated they were about to speak, but who lost the floor to people at the other site. Those at the other site interpreted the momentary pause, slightly longer than customary in face-to-face situations, as lack of inclination to speak. Participants must learn to speak deliberately, with obvious signals when finished and a space break before the next person takes up the conversation. Exaggerating normal conversational cues is not ideal, though, since so much meaning is conveyed through hesitation, rhythm, and timing.

There was rarely a feeling of equal partnership in video team meetings: one site or another always had more people or more dominance, as though they were having a meeting and the other site was dropping in, trying to get what they could out of it. At its worst, the two sites would give up and begin each talking among themselves, often in different languages, until someone would take the lead to pull the conversation back to a dialogue.

These problems can be alleviated if not obviated. A group facilitator can remind the team of how communication must be different when mediated via video, and the facilitator can break in to comment on why certain people are not being heard or to call attention to miscommunication. Team members can be coached on their verbal style, and the team can engage in direct conversation about how people interpret tone of voice or how turn taking is normally signaled, so that the team members can use such metalinguistic awareness to work toward better meetings. Because many team members are not comfortable with technology, the facilitator can make sure the cameras and mics are adjusted and that technical assistance is brought in when necessary.

A consultant/facilitator with a background in professional communication, one who understands group process and conflict, conversation structure and effective meeting organization (and who is comfortable with communication technology) can provide significant benefits to companies who want smoothly productive video meetings. If teams become accustomed early in their development to talk directly about team communication in both face-to-face and mediated situations, then raising issues of how other team members communicate, or fail to communicate, is made easier down the road. Such discussions can include cautious observations about cultural and linguistic differences, realizing that individual differences are likely to contribute as much variation as those attributable to group membership, and that any attempt to generalize on the basis of nationality risks stereotyping. Team leaders can also be coached to pay attention to team dynamics, to invite quiet participants to speak and to speak up, and to alter the pacing of conversation to increase participation across language groups.

Whiteboard Software

Videoconferencing has serious limitations for document-intensive work, as for prototyping reports or reviewing long and complex documents, and must be usefully augmented with other collaborative technologies. Whiteboard software allows teams to prototype documents or develop ideas, using a single computer screen or linked computers to create a shared space for visual thinking. The idea is fairly simple (not much more than a flipchart, really) but quite powerful in its effects on meetings. Instead of merely meeting to talk, a team with whiteboard software can meet, talk, *and* see their words (and work) displayed. Making thinking visible helps teams see what they say and thereby agree on what they think and will do.

Several companies market whiteboard meeting software, including LiveWorks™ (a spinoff from Xerox; point your net browser to http://www.liveworks.com), which has a program called MeetingDesk™ that will work on a PC or displayed on LiveWorks' large format LiveBoard (see below). Fujitsu markets Desktop Conferencing™ (DTC) software, which includes a whiteboard function, and Microsoft has a new product, NetMeeting™, that supports whiteboard functions in Windows 95™. In essence, whiteboard software offers a dynamic surface for writing freehand with an infrared light pen directly on a screen or palette. Whiteboard software answers a traditional criticism of PCs, that they are good for writing, but not very good tools for brainstorming and planning because they lack the graphic capability of a blank sketch pad.

Whiteboard software allows handwriting to float in the public space of a computer display. It supports invention and planning with built-in features that allow words and phrases to be selected, moved, ordered, copied, or deleted. Because it is not restricted to typed input but supports free pen movement, a whiteboard allows groups to develop diagrams or flowcharts, mindmaps, storyboards, or doodles. It provides a more dynamic space than a flip chart, in that "pages" can be as long as one likes, and multiple pages can be sequenced to reflect the group's work and accessed via an index or "go to" function. Whiteboard software is superior to flipcharts in its revising capabilities, so that during a meeting a team can move from brainstorming to ordering ideas and emerge from a meeting with a well ordered document prototype, development matrix, or working plan. Vendors of whiteboards have partially addressed the limitations of flip charts by creating pressure sensitive whiteboards with built-in printing facility. But for flexibility and ease of revision, it is hard to beat whiteboard software running on a computer.

Whiteboard software works equally well as a stand-alone application for a single-site meeting as for meetings that bridge two or more sites with linked computers. As a stand-alone tool, whiteboard software offers the functionality of a blackboard (a familiar thinking tool with a revered history), with the added virtue of unlimited space and ability to print at the close of a meeting. Using whiteboard software is relatively easy, though it does suffer from the same kinds of disorientation common to computer users who must view large scrolling texts through the small windows provided by an operating system and monitor.

I helped facilitate a day-long meeting of preclinical scientists and regulatory experts to review drafts of an expert report, a document that critically assesses a company's development program. We worked throughout the day on the computer whiteboard, generating about thirty pages of notes for the three authors, detailing what they were doing well in their sections and what they needed to attend to. We identified key messages and issues to be incorporated. We also prototyped introductory and concluding discussion sections, and identified page lengths for each part of the document. We used a series of whiteboard pages to list issues that needed scientific discussion and resolution outside the review meeting, a timetable with dates for the next draft cycling and review session, and a list of action items involving other team members (links to critical issues in clinical development and items to take up with the FDA at the next meeting).

The whiteboard file we generated was large, about 1.6 MB, but we were able to print it and distribute it at the end of the day. All during the day, the group's attention was focused on the whiteboard, ideas were refined, work was accomplished and documented in a public space for all to see and agree to. The efficiency gains were significant over meetings that are largely undocumented talk, where everyone would leave with their own interpretations of what was agreed to. And nobody had to gather up fifteen flipcharts at the end of the day, try to make sense of everything that went on, type it up, and distribute the notes.

Document Projection Systems

There are several technological solutions to creating shared visual working space. Desktop computers can simply be linked to each other, so that users share the same screen while they sit at their own desks, sometimes with small video cameras mounted on top of each computer so a low resolution video picture-in-monitor is projected across computers. However, since one of the goals of our intervention was to get individuals away from their desks and to collaborate in team spaces, we looked for displays that could accommodate team viewing.

One option that works fairly well is to link the computers to computer projectors, preferably not the flat panel LED displays that work in tandem with overhead projectors, but units that contain their own projection systems and that project the VGA output from a computer. These portable units (which retail for somewhere between $4,000 and $12,000 from such companies as InFocus and Polaroid) have been much improved in recent years in both portability and affordability, and they can project truly high quality screen images even under normal ambient light. Conferencing software can join remote computers, so that more than one site can see the same image, and pen palettes support working with a light pen, which is much easier than attempting to draw with a mouse (though it is best to have both input devices).

Even better than a projector is a product that we used from LiveWorks, called the LiveBoard™, running MeetingBoard™ software, the parallel software to MeetingDesk™ (which runs on a desktop PC). The LiveBoard™ is a large unit that contains a PC, a large, rear projection screen, and a light pen. Anything you can do on a PC, you can do on a LiveBoard™, only it is displayed in high resolution with good readability in an image that measures 67 inches diagonally. LiveBoards™ can be connected via a multipoint bridge, so that people at multiple locations share the same computer screen. The company has built some nice features into the LiveBoard™: a lightpen interface with limited character recognition, Olé (Object Linking and Embedding) so that application files can be embedded in MeetingBoard™ files, some capabilities for bringing in document snapshots, and flexible list creation and comment tools. The shared space itself is the real advantage, however, since team members at different sites can view and contribute comments to the same document space.

The LiveBoards™ are tremendous tools for allowing globally dispersed teams to conveniently plan, develop, and workshop complex documents across sites. There is a deep qualitative change in the ways teams work during meetings once their group intelligence is focused on a dynamic display, where their thinking, decisions, and language can be recorded and reordered. During electronically facilitated meetings, our teams were able to work up and rehearse presentations, display and annotate PowerPoint™ slides, prototype scripts, and offer critical commentary to presenting team members. Our teams regularly prototyped documents, reviewed and annotated drafts, and engaged in careful discussion about whether their documents delivered clear messages in prominent positions (with sometimes extended debate on such burning issues as whether it is appropriate to bullet key findings in a scientific report!). Just prior to filing, the teams used the technology to review various documents for consistent delivery of messages and treatment of issues, using the LiveBoards™ to record necessary links and specific language that should be consistent throughout the dossier. All of these tasks are document intensive, and stand in contrast to videoconferences featuring talking heads.

Conferencing Software

Whiteboard software is one of several kinds of conferencing software, designed to work on linked computers so that users at several sites share the same screen. With our teams, we made extensive user of MeetingBoard™ for the LiveBoard™, which allows a Word™ or PowerPoint™ (or other) file to be printed to disk and then displayed as a bitmapped file on the LiveBoard™, where the file can be annotated using MeetingDesk™ tools. In essence, we were able to make notes on top of the image of the file (but not edit the file itself).

What the LiveBoard™ didn't have, and what our teams really wanted, was true application-sharing software, so they could run standard Windows™ applications and work on actual document files during meetings. The MeetingDesk™ designers imagined, logically enough, that a bitmapped image of a document would suffice for annotating, marking trouble spots, suggesting corrections, and so on. But much of the work of team reviews involved very detailed discussion of complex messages and supporting responses within highly complex documents. The teams wanted to enter a review meeting ready to debate document issues. They needed to have before them a dynamic representation of the actual file, so that as substitutions, deletions, or additions were suggested, they could see the new language on the screen and agree that the new language was indeed what was needed in the document. Seeing annotated text is never the same as seeing the actual text in context.

True application sharing is demanding in terms of system resources and operating system functionality. The need, however, is well recognized, and the software market is integrating such functionality with the latest operating systems (Windows 95™ and Windows NT™). Fujitsu Desktop Conferencing™ is designed to support application sharing, as is Microsoft's NetMeeting™. LiveWorks™ now has a Windows 95™ version of its software with application sharing.

While many other kinds of groupware are available, for the most part designers have not made the assumption that groups need to work intensively on complex documents in real time. Rather, much groupware is founded on the presupposition that the goal is to model or facilitate oral behaviors typical of face-to-face meetings: brainstorming, listing, building consensus, and ranking or eliminating suggestions (see [14] for an assessment of the empirical research). Synchronous discussion is supported by groupware such as IBM's Person-to-Person™ and classroom software such as Daedalus™, and the nature of such conversations is well investigated in the literature of computers and composition. But because phone and videoconferencing are so ubiquitous and familiar in pharmaceutical workplaces, there is little perceived need for chat facilities. One merely picked up the phone and dialed a conference call or reserved the video-conference facilities. There may be benefits of realtime talk via computers, but they remained unrealized in the pharma workplaces where we were involved.

Shared Drives

Shared drives from central servers allow team members to have access to current project files on an anytime/anywhere basis as long as they have sign-on privileges. In our intervention, the shared drive allowed team members to access current versions of files, pull them up on the LiveBoard™, at their desks, or from affiliate sites when team members were on the road. We organized separate drives for each development team, passworded so only team members could access the files. Uses of the shared drives grew over time, with teams organizing files for their presentation scripts and slides, frequently asked questions, development matrices, strategic documents, meeting notes, MeetingDesk,™ files, budgets, timelines, prototypes, and NDA documents.

Using a shared drive raised concern among team members about who had access to what files, at what times, and for what purposes. Shared drives make information public within a company, and there are always a variety of reasons for not sharing information. Some authors simply did not want their files in a public place until they felt comfortable that the files were ready (final draft stage). Others did not want early drafts of information on the shared drive because they thought there might be inaccuracies that would find their way into other documents. Because the drives had open directory structures with open access to everyone on the team, files could have been subject to malicious or inadvertent damage, though there was never evidence of such.

Significant indirect learning was an unexpected benefit of the move toward shared drives. Team members learned routines to log onto remote drives and navigate directories set up by others. Reading and writing to remote drives revealed machine and network bottlenecks, which led to technical changes and overall improvement of the network efficiency.

The shared drive also raised problems of creating directory structures for large groups of documents, including prototypes, general information files (schedules, directions, memos), and data files. Just figuring out naming conventions was a challenge within the eight-character name stricture that is a DOS legacy. All the chaos that typifies the hard drives of many individuals is suddenly amplified by team members who come and go, leaving discarded files and versions of files, temporary files, and just plain garbage files in their wake. The best-laid plans, the carefully set out directories, the guidelines on file naming conventions—all mean very little in the hurly-burly of people rushing about to get work done. These difficulties with shared drives exemplify the more general challenges of the transition from the computer as a tool for personal and individual work to the computer as a tool for social interaction and teamwork. While some teams gave up in frustration or never really made good use of a shared drive, others quickly took to the tool and came to rely on it and own it as a tool to enhance teamwork.

We have a long way to go in developing protocols for working in shared directory spaces; in deriving a logic of file structures that will seem logical to diverse, creative minds; and in finding ways to train or support new users who are intimidated by moving from personal computing to social computing.

Revision Software

Word processors that offer revision mark-up features provide a relatively simple groupware tool with important implications for how teams handle document review and revision. Our client had standardized around Microsoft Word™, which allows files to be protected for revision marks and passworded to the author, meaning that the original file is preserved, but reviewers can annotate, change, or delete right on the text. Word™ marks all suggested changes and the author can then review and accept or reject any suggested changes. Up to eight reviewers can mark up a text, and Word™ keeps track of who suggested what changes when with color-coded revision marks.

With files on a shared drive, passworded to the author, and protected for revision marks, team members could be notified via e-mail that a draft was ready for review on the shared drive, and team members could open the file, indicate various revisions right on the text file, and save it to the shared drive. The next team member to open the file could see the original text and the first reviewer's changes and remarks, and then add additional commentary and contribute to the dialogue among the author and other reviewers. On one team, we followed this procedure with the prototype and developed drafts of the package insert for the drug, with three authors contributing sections, several team members posting commentary, authors reviewing the comments, and then meeting to review the proposed revisions and discuss further changes. The process made document development and review a public, team-centered process, with authorship and ownership of the document passing from individual control to the team.

Making reviews public changes business as usual. The typical process would have an author attach a file to an e-mail message to all reviewers and then integrate the comments and suggestions. Reviewers would be unaware of other reviewers' comments, and the author would have full control over responding to suggestions. The combination of keeping files on a shared drive, having the comments of various team members recorded with revision marks, and then meeting to review the marked up text meant that the control that would have been centered in the author was now distributed across the team. This change, toward team ownership of documents, was alternately welcomed and resisted, depending on who the key players were for a given document and how far they endorsed collaboration.

BRINGING THE TECHNOLOGY TOGETHER

All of these team technologies can come together in a physical and virtual space in the form of a *team room,* where the technologies are clustered and the room is designed to facilitate global work. Team rooms can be augmented with printing and faxing, with direct phone connections and personal computers, all tied into a network of rooms where teams meet and work. Such a room can be built to accommodate from four to sixteen people, with tables arranged in a semicircle oriented toward the video cameras and the large display, either a projected image of the computer screen or a LiveBoard™. Shelves can hold printed resources and eventually the team's output of documentation, in draft or signed-off versions, and tracking charts or other visuals that convey the team's positioning and accomplishments can line the walls. The room can be a place not just to meet, but to work. The concept is similar to a war-room that is sometimes set up for major proposal initiatives, but with full global connectivity. The connectivity can be extended to team members' individual workstations, so they can join the working space on the spur of the moment, as their participation is needed for a meeting, or when it makes sense to meet desktop-to-desktop.

Such team rooms will not come cheap, though the growth of the net is forcing the cost of high bandwidth communication down and, as we all know, more computing power is available for less money each year. A cost estimate of $200,000-400,000 is within range, with something on the order of $200,000 per year to run one. That is a significant investment, though it is worth remembering an earlier remark, that the top 100 drugs at peak sales generate one million dollars per day, with blockbusters generating $2 to $3 billion per year. Technologies that speed time to market can be rationalized around those numbers.

NEW FORMS OF TEXT

New forms of text are emerging in the pharma industry. For example, *hypertexts* are being constructed to supply the health authorities with electronic copies of new drug applications. The primary objective is to provide data files so the authorities can run their own statistical analyses of the data on safety and effectiveness, but it doesn't take a lot of imagination to see advantages in delivering new drug applications in electronic form. As in other settings, the initial electronic files are simple dumps of the data tables, with a few pointers and links. But hypertext holds real promise for the industry, and companies are moving quickly to provide hypertexts that provide efficient tools and structures for visualization and manipulation of the data. A new drug application represents years of time and millions of dollars invested in what one hopes is a coherent research and development process, and the application must show links among the studies, within the chemical and manufacturing processes, and among the various arguments made in various sections of the document. Graphical browsers, nested information, layered structures, glossaries, bookmarks, menus, indexes: all are usefully beginning to inform hypertext presentations. It is quite natural to hear the medical researchers, toxicologists, and chemists talk about *links* among various documents, and while what they have in mind is not a dynamic hypertext link, but a textual reference across volumes, the notion is the same and the technology holds promise as a medium suited perfectly to the messages that must be delivered.

Document databases will increasingly organize the huge amounts of information and complex sets of related documents occasioned by NDAs. In the pharma industry, Documentum™, from the company of the same name, another Xerox spinoff company, is the leading document database software (for a description of its use at Glaxo Wellcome pharmaceutical, see http://www.documentum.com/glaxo.htm). Documentum™ in tandem with publishing software such as Xerox Document Assembler™ (XDA, described at http://www.xerox.com/XPS/newpage/nrxda.htm), allows for various levels of document production and control, insuring accurate data transfer from one document to another, consistent format across documents written by many different authors, and high-level coordination of versions and links among complex documentation sets (indexing, tables of contents, pagination, links across volumes). Such systems have check-out systems for documents with various levels of access and authority over documents. Once a document routing slip is filled out, Documentum™ will send out e-mail notices to reviewers that a text is coming their way, route the electronic text in and out, keep track of versions and revisions, allow annotation and text markup, remind reviewers when their comments are due or overdue, and record signatures on documents that need signed approval.

The thrust of software development has been integration across platforms and applications, so that a company with a mix of hardware and existing expertise with publications software such as InterLeaf™ can preserve and leverage that experience. Like any software system, the logic of any chosen piece of software will to some extent run counter to the logic of work established by an organization or a team, and some refitting of the software to the organization or organization to the software will be necessary. It is not an assumption of Documentum™, for example, that anyone other than the author is interested in seeing the comments of reviewers, but as noted above, there is good reason for encouraging dialogue within review cycles.

Increasingly, teams will be aided by shared intelligence via *intranet sites.* Opper and Fersko-Weiss [8, p. 26] model groupware around three essentials: electronic communication, a group focus, and management of shared information. Such tools as shared drives, document databases, and more recently intranets, play key roles in information management. Teams need access to current versions of reliable information, information that is typically widely spread around an organization. Intelligent systems, or expert systems, navigated with Web browsers and managed with document databases, are quickly emerging as key information-support tools. Sites can be constructed for internal use, or with areas passworded to provide or deny access to individuals within and outside the company. File management tools such as Documentum™ provide the server structure for intranet development with clients accessing files with their net browsers.

Hybrid intranets can allow team members to access reference literature on regulations, discuss issues with or post questions to other team members or to experts throughout the organization via bulletin board functions, look up files or versions of reports, and enter the Web for access to sites on pharmaceuticals, diseases, and healthcare. The WWW will increasingly be a place where information about drug products and disease management is found—a place for support groups, prescribing physicians, and disease sufferers to find the information they need. Pharma companies are rapidly migrating to the WWW for many purposes.

IMPLICATIONS FOR PROFESSIONAL COMMUNICATION PROGRAMS

Team communication needs in the pharma industry are complex and demanding. NDAs are large, complex objects, and they represent just one example of documentation produced by the industry. When the needs are to work across sites, to access the documents on an anytime/anywhere basis, and to engage in document-intensive tasks such as application sharing, the network, hardware, and software resource demands are substantial.

Applications of technology to document-intensive group tasks push the leading edge of technology development. The information-technology staff within pharma companies and even the technology vendors themselves are only beginning to understand why and how to make technology work in such highly social and highly textual ways. Computers were not developed for such purposes. They evolved from data-manipulation machines, morphed into personal productivity tools, but have only recently been reconfigured to support socially structured work. They only awkwardly meet the challenge of work that centers on massive documentation tasks. But each new operating system enhances network functionality in support of group computing, and the future promises expanded connectivity and resourcefulness.

We are only now discovering how we need to work with global teams to produce complex documentation sets. How should a team review be structured when the documents are complex and the team members dispersed around the globe? How can team members be encouraged to communicate effectively across the distances and the separation imposed by new media? What are the best ways to develop documentation so that a huge submission is actually a carefully orchestrated, coherent, message-driven issue-oriented dossier? Each cautious step toward implementing new systems demands new user knowledge, new mental models for how computers enable work processes, bigger and faster hardware and networking, and significant technical support for assimilating and accommodating the designed software to existing work processes.

I think it is worth pointing out some of the more obvious ways that our practices in professional communication (and, indeed, in composition teaching generally) can help prepare students to contribute to the processes of document development through social technologies within industry. Within our programs, as we think about preparing our students, we might consider whether we build and reinforce the following kinds of understanding and skills:

- **Writing should be developed as a tool for visual thinking.** Students can use writing as a learning tool, to capture and articulate what they know and share. Students can come *to see what they know* by arranging ideas within something resembling a messages and issues table and transforming the table to an early document prototype©. Technology enhances visual intelligence and provides a visual space for creating and manipulating ideas.

- **Writing is best treated as a process.** Good writing is best created by dividing work into stages in a managed process, from conceptualizing via a messages and issues table, to creating an initial document prototype©, to revising the prototype and advancing it toward increasingly mature drafts, to taking drafts through several rounds of focused revision, to publishing for various markets. As exemplified in this chapter, prototyping around messages, issues, responses, and support can be a powerful heuristic during the planning stage of documents. Technology can enable the process.
- **Professional communication classes should integrate writing, speaking, and working.** Through interdisciplinary studies, our students should be encouraged to develop their understanding of leadership, teamwork, and communication (small group, cross cultural, organizational, and technologically mediated). They should study project management and engage in the production of document libraries so they understand the dynamics of large documentation projects.
- **Writing must be understood and practiced as social and collaborative.** Cross-functional teams attain a synergy from involving people of different expertise, experience, and cultures in a common task. At every opportunity, students should be encouraged to work in teams, to contribute differentially based on their expertise, to read about teamwork, to experiment with various forms of leadership and work structures, and to reflect on their experiences as members of teams. Whenever possible, technologies should be exploited for their support of collaborative interaction.
- **Technology should be developed as a social tool.** We should use shared drives, e-mail, soft-copy review, and mark-up systems to explore the ways that people can work collaboratively across a network. We should use technology to bring writing into public space and to collect distributed knowledge in a shared space. Many tools (such as revision marks) are available but unavailed within our classes and our work; we tend to ignore the ready-to-hand tools embedded in our familiar applications as we seek some powerful new form of groupware. We would do well to rethink the tools we have in the interest of helping students and ourselves develop social uses of technology.
- **Technologies should be used to bridge distances.** We should pursue class collaborations across universities and with industry, even if the results at times tend toward chaos. Dealing with unexpected contingencies, technological breakdowns, and imminent disaster is perfect preparation for the modern workplace. We should encourage our students and become involved ourselves in working with distance education, virtual classrooms, and video and intranet delivery of courses. At the same time, we should research the effects of various technologies on group communication.

- **Electronic texts in various forms should be explored as a rapidly expanding alternative to paper-based texts.** All students, including those in technical communication, have an unparalleled opportunity to become information managers and designers of new forms of hypertexts, Web sites, and document database systems. In all our classes, we should be moving back and forth between paper and online sources for research, writing, and communication. We should participate in designing technologically rich environments that accommodate the special demands of working with complicated texts in electronic space, using writing classrooms as our labs.

Many writing programs in our colleges and universities are already actively engaged with the sorts of learning outlined above; their students will discover useful roles in industry settings for technical and professional communicators. The challenges are great and intensely interesting, since the work brings together the document-development practices of rhetoric and technical communication with the social dynamics of globally dispersed teams, all within an environment constructed out of the complexity of emergent technologies. We should, above all, encourage our students to think imaginatively about the intersection of work, technology, and communication, and to be prepared for a workplace that will change faster than we can imagine. Our students' resourcefulness, their agility, and their willingness to work creatively within technical environments will ensure a valued place for them in the workplace.

REFERENCES

1. C. J. Forbes, The Role of the Technical Communicator within Organizational Information Development Cycles, in *Publications Management: Essays for Professional Communicators,* O. J. Allen and L. Deming (eds.), Baywood, Amityville, New York, pp. 95-106, 1994.
2. J. Prekeges, Planning and Tracking a Project, in *Techniques for Technical Communicators,* C. M. Barnum and S. Carliner (eds.), Macmillan, New York, pp. 79-106, 1993.
3. B. Weber, Project Management: The Art of Managing Deadlines, in *Publications Management: Essays for Professional Communicators,* O. J. Allen and L. Deming (eds.), Baywood, Amityville, New York, pp. 107-116, 1994.
4. B. Couture and J. Rymer, Interactive Writing on the Job: Definitions and Implications of Collaboration, in *Writing in the Business Professions: Research, Theory, and Practice,* M. Kogen (ed.), National Council of Teachers of English, Urbana, Illinois, pp. 73-93, 1989.
5. L. M. Jessup and J. S. Valacich, *Group Support Systems: New Perspectives,* Macmillan, New York, 1993.
6. R. Johansen (ed.), *Groupware: Computer Support for Business Teams,* Free Press, New York, 1988.
7. P. Lloyd (ed.), *Groupware in the 21st Century,* Praeger, Westport, Connecticut, 1994.

8. S. Opper and H. Fersko-Weiss, *Technology for Teams: Enhancing Productivity in Networked Organizations,* van Nostrand Reinhold, New York, 1992.
9. J. Lipnack and J. Stamps, *Virtual Teams: Reaching across Space, Time, and Organizations with Technology,* Wiley, New York, 1997.
10. M. Shrage, *No More Teams! Mastering the Dynamics of Creative Collaboration,* Doubleday, New York, 1995.
11. T. K. Bikson and J. D. Eveland, The Interplay of Work Group Structures and Computer Support, in *Intellectual Teamwork,* J. Galegher, R. E. Kraut, and C. Egido (eds.), Erlbaum, Mahwah, New Jersey, pp. 245-289, 1990.
12. C. J. Hansen, Contextualizing Technology and Communication in a Corporate Setting, in *Nonacademic Writing: Social Theory and Technology,* A. H. Duin and C. J. Hansen (eds.), Erlbaum, Mahwah, New Jersey, pp. 305-324, 1996.
13. C. Egido, Teleconferencing as a Technology to Support Cooperative Work: Its Possibilities and Limitations, in *Intellectual Teamwork,* J. Galegher, R. E. Kraut, and C. Egido (eds.), Erlbaum, Mahwah, New Jersey, pp. 351-371, 1990.
14. K. L. Kraemer, Technology and Groups: Assessment of the Empirical Research, in *Intellectual Teamwork,* J. Galegher, R. E. Kraut, and C. Egido (eds.), Erlbaum, Mahwah, New Jersey, pp. 375-405, 1990.

CHAPTER 4

The Impact of Discourse Communities on International Professional Communication

JOHN WEBB and MICHAEL KEENE

Our discussion in this chapter is based on an analysis of data obtained by the National Aeronautics and Space Administration (NASA) in the NASA/Department of Defense Aerospace Knowledge Diffusion Research Project (henceforth "the Project"). The Project was begun in 1989 to examine the flow of scientific and technical information at the individual, organizational, and international levels. A joint effort of the Indiana University Center for Survey Research and the NASA Langley Research Center, this multi-phase Project is providing data that should prove useful to information managers, research and development managers, and others who are concerned with improving the utilization and communication of scientific and technical information [1].

One of the Project's goals is to develop a clear explanation drawn from the professional engineering community of what constitutes "acceptable and desirable communications norms" within that community. The Project thus collects descriptive data concerning the nature of use, frequency of use, and importance of technical communications to engineers in the workplace. In Phase 2, the Project surveyed 2,355 aerospace engineers and scientists in the United States on these topics. The Project's Phase 4 cross-national surveys, from which the data discussed in this chapter are taken, were designed to help the U.S. aerospace industry understand the diffusion of knowledge in the aerospace industries of other nations. Phase 4 activities correlated the survey responses of aerospace engineers, scientists, and academicians in India, Japan, the Netherlands, Russia, and the United States on questionnaires very similar to the one directed at U.S. aerospace professionals in Phase 2. This chapter's first goal, then, is to provide a comparison of the responses of aerospace professionals in these four foreign countries with the responses of U.S. scientists and engineers to the Phase 4 survey questionnaires (*n*, the total number of professional respondents in Phase 4, = 823).

In a previous analysis of single-country survey results for the Project, emphasis was placed on the importance of understanding a foreign culture—in that case, Japan—when interpreting empirical data that have been collected in that culture [2, p. 71]. In order to ground that reasoning in a comprehensive theoretical framework, a subsequent examination of the reacculturation of aerospace engineering students to the world of work based on the Project's Phase 2 (U.S. professional) and Phase 3 (U.S. student) data adopted Bruffee's [3] description of the socially constructed nature of human knowledge as the "higher-level theoretical framework . . . within which the interpretation of such data can take on consistent and fuller meaning" [4, p. 493]. Within such a culturally sensitive, social constructionist framework, in our own analysis of the Phase 4 data for another publication [5], it clearly seemed necessary to discuss how language, culture, and sociopolitical conditions in each of the four nations—India, Japan, the Netherlands, and Russia (not to mention the United States)—might affect the technical communication practices of aerospace professionals in that country. How else might one determine how knowledge is constructed and disseminated in an international field such as aerospace engineering and science? The very complexity of this effort—comparing the languages and cultures of five countries—is what initially led us to question its overall significance to a meaningful analysis of the Phase 4 cross-national data. For example, India alone has twenty-five major language groups used by some considerable percentage of its population, four major religions and a number of minor ones, and abundant regional distinctions [6; 7, p. 67]. Thus, a portrait of sociopolitical, cultural, and linguistic impacts on the communication practices of Indian aerospace scientists and engineers could only be painted in the broadest strokes, which we did in our earlier article. Portraits of Japan, the Netherlands, and Russia evoked different problematic aspects in this complicated montage.

As we wrote about how language and culture in each country affected aerospace professional communication in that country, as we tried to update our discussion to reflect sociopolitical flux, and as we tried to acknowledge the important ways each community's language in fact created and reshaped the communities in question, our frustration grew. In light of the fact that we were analyzing data on communication practices gathered from participants in a predominantly English-language industry that oversteps international boundaries and transcends political affiliations, the question perhaps was inevitable: how much do we have to know about a particular country to understand the discourse of its aerospace scientists and engineers when most of that discourse is in English?

That question led us to this essay's further examination of the Phase 4 data, here within a reconsideration of the concept of "community" as commonly used in discussions of communication. Pursuing that reconsideration is our second goal.

The next section reviews the literature relevant to our reexamination of the data within this context. Following the literature review is a summary of findings from the Phase 4 survey responses of aerospace professionals in five countries and a discussion of the significance of these findings to international communication among aerospace scientists, engineers, and academicians. In our concluding remarks we discuss the broader implications of our findings for the study of international professional communication.

LITERATURE REVIEW

Mastering international technology transfer and information flow is more important than ever to the aerospace industry. Federal government sources point out that increasing international cooperation between U.S. and foreign companies in air and space projects is a continuing trend [8]. Multinational aerospace projects, such as the Boeing 777, have become commonplace, and the documentation for such projects is usually in English. According to the official history of the Federal Aviation Administration, in 1952 the International Civil Aviation Organization (ICAO) adopted English as the official language of aviation after the United States agreed to drop its version of the phonetic alphabet (able, baker, charlie, and so forth) and adopt the internationalized version (alpha, bravo, charlie), a concession which related to the sensitivities of the other English-speaking countries [9]. The official journals of the ICAO are published in English, whatever the native country of the authors. European Space Agency (ESA) documents can be written in the language of the country of origin, but the common language of scientific communication among ESA members is English. Aerospace documents of the North Atlantic Treaty Organization's Advisory Group for Aerospace Research and Development (AGARD) are published in both English and French. In a recent analysis of the Project's Phase 3 (U.S. student) data that compared the responses of survey samples made up of native-English-speaking and English-as-second-language (ESL) student members of the American Institute of Aeronautics and Astronautics, more than 75 percent of the ESL sample reported that they spoke English fluently, and 90 percent reported that they read English fluently [10]. English is thus the official as well as the practical language of aeronautics and astronautics.

Of the countries whose aerospace engineers and scientists make up the survey samples described in this chapter, only Russia does not use English as the collective tongue of technical and professional communication (and that situation may be rapidly changing). In India, a nation of many distinct language groups and diverse cultures, colonially derived English has unified the curricula in schools and universities and is the national and international language of science and commerce. In Japan, Stevenson [11] reported that in a survey of seventy engineers and managers 44 percent of those interviewed wrote in English only, 22 percent in Japanese only, and 34 percent write in both [2, p. 67]. Bailey and Gorlach [12], Chesire [13], Greenbaum [14], and Kachur [15] document several aspects of the use and adaptation of English in India, Japan, Ghana, and other countries [16]. Berns predicts that it is "likely that English will become the primary language of citizens of the EC (European Community). Whether or not it is ever officially declared such, it will be even more widely used as a vehicle for intra-European communication across all social groups" [16, p. 9]. The cosmopolitanism of the Dutch and their historic, continuing prominence in international trade occasion a high rate of anglicization in the Netherlands [17]. "In 1990, the percentage of Dutch adults who speak English was 68%, which compares with 100% in the UK, 99% in Ireland, 51% in Denmark, 43% in Germany, 26% in both Belgium and France, and 13% in both Italy and Spain" [18, p. 49].

Because English is emerging as the worldwide common tongue not only of aerospace technology but also of professional communication in general, it has become a commonplace in professional and technical communication that we need to be especially aware of the cultural differences inherent in communicating with nations other than our own, native-English-speaking nation. This need for sensitivity to difference is further underlined by the fact that English as it is known in the United States may not be exactly the dialect of English that evolves as the common professional tongue. According to Dennett in "World Class Status Does Not Ensure World Class Usage" [19], "English may be the language of the global village, but the villagers are far from agreement on what is good use of the language—particularly the written language. In fact, 'The world may be turning into a global village, but the villagers are at odds over what constitutes good writing' [20]." Dennett also quotes a summary of Subbiah's [21] argument: "people from different nations think differently, learn differently, and write differently; their notions of good technical English do not necessarily have to be the same as those of native speakers of English."

Subbiah's argument reflects the current emphasis on "multiculturalism"—by which is meant "a diversity of cultures sharing power and wealth" [22] in the workplace and in academic curricula. Zee asks, along with Schechner [23], if the dialogue on multiculturalism should not be about "interculturalism"—by which is meant "the active exploration of the contradictions, problems, faults, and possibilities of multiculturalism." Otherwise, in trying to bring a multicultural perspective to understanding international technical communication, we may in fact be doing nothing more than proselytizing our own culture. By culture, we mean "the rich social life that constitutes a people, their customs, styles, tastes, festivals, rituals, gods" [24, p. 187]. Culture is "an established set of values and a way of thinking and behaving that is passed from generation to generation" [25, p. 53]. Specific to our need in the United States to recognize and understand cultural differences, culture "is an awareness individuals have, or do not have, when they meet people from different nations" [21, p. 14]. For residents of the United States, "it is as if our culture is the way of life toward which everyone else is striving . . . everybody's actions have the same meaning and arise from identical motivations" [26]. As these comments show, the ascendancy of English as the world language of professional communication actually increases our need for understanding other cultures whose citizens may also use English for their professional and technical communications.

On the one hand, using English as common tongue is, above all, a practical affair. In "Language Choice and Cultural Imperialism: A Nigerian Perspective," Bisong [27] concludes that

> although English is the official language of Nigeria, it has not succeeded in displacing or replacing any of the indigenous languages. It performs a useful function in a multilingual society and will continue to do so, since no nation can escape its history. But attitudes to the language have changed since colonial times. It is no longer perceived as the Imperial tongue that must be learned at all costs. Reasons for learning English are now pragmatic in nature, and run counter to Phillipson's argument that those who acquire the language in a situation where it plays a dominant role are victims of linguistic imperialism [28, p. 131].

Thus Dakich interviewed a cross-section of international engineers not unlike the survey sample of aerospace engineers and scientists described in this chapter, who observed that most of their English instruction focused too heavily on structured learning of syntax and grammar [29, p. 565]. And McDaniel [30, p. 31] cites Yamada that "one commonly held belief is that cross-cultural misunderstandings [may] result from a nonnative speaker's inability to generate nativelike phonological and syntactical constructions" [31, p. 13]. So McDaniel concludes that "the view that nonnative speakers writing in English ought to master its logic ignores the cultural complexity of language" [32, p. 571].

On the other hand, in communication theory today, language is considered not just a function of culture, but rather language, like knowledge, is thought of as a construct of culture. "Because language is a manifestation of culture, differences among cultures are reflected in the language of various cultures" [33, p. 553]. Uljin and others, however, found in a quantitative, text-based experiment in the Netherlands that understanding documents is not culture-dependent: "universal writing rules should be applied" [34, p. 573]. Uljin and his fellow Dutch researchers found that "a 'culturally' rewritten technical manual does not perform better than a translated one. However, a rewritten technical manual is preferred when users are asked for their preference. Culture does not significantly affect document composition" [34, p. 577]. In a qualitative survey of Asian and American editors and students, Campbell and Bernick indicated that "documents might . . . be most effective when they employ the rhetorical strategies of the language and culture that they are written in, rather than trying to anticipate a reader's cultural expectations" [35, p. 42].

It is not our intent here to argue that knowing about the culture of someone with whom you are having a conversation or an e-mail dialogue or a business transaction is unimportant to understanding that exchange. Distinctions in the cultural processes of thought, learning, speech, and writing must indeed be taken into consideration, for example when data such as that from the four foreign nations presented in this chapter are compared to the data for the United States also reported here. We still endorse the view most often expressed in technical and professional communication today, that the "higher level, theoretical framework" [4] for interpreting data such as that presented in this essay is that knowledge is a construct generated "by communities of like-minded peers" [36, p. 774]. Our intent, rather, is to contest the notion that adoption of Bruffee's social constructionist model necessarily entails exhaustive exploration of every culture we wish to study in the kind of cross-cultural analysis undertaken here.

Bruffee [37] and other researchers adapted this model for the study of communication and composition based in part on Thomas Kuhn's [38] model of a scientific community in which Kuhn articulated, among other perceptions, the concept of "paradigm shift." Longo [39] explains Kuhn's scientific community in this way: In Kuhn's model, the community of scientists create a paradigm—they share a world view, values, assumptions, *language* (our emphasis), and a way of legitimating knowledge—and there is consensus within the community. Within this consensus, the scientists identify problems to work on and gain knowledge from the work. This is what Kuhn calls "normal science." The consensus is lost when new world views enter the community, as problems emerge that the existing paradigm cannot account for, and a new paradigm gains support. As the new paradigm is legitimated as knowledge, the problems the community works on shift away from problems routinely identified and dealt with under the old paradigm—Kuhn calls this shift "revolutionary science"—and knowledge within the community changes [38, p. 148].

Longo also posits that in adapting Kuhn's model to communication and composition studies, an adaptation in which Kuhn's "scientific community" becomes Bruffee's "community of discourse," Bruffee and the other communication and composition researchers have chosen the conventions of discourse they study by focusing on what Richard Rorty [40] called "normal discourse" (like Kuhn's "normal science") to the exclusion of "abnormal discourse" (analogous to Kuhn's "revolutionary science"):

> But this model of discourse community presupposes an autonomous community in which people within the community do not or cannot communicate with people outside the community. Moreover, this consensus-based community engaging in conventional discourse does not harbor different opinions among the people within the community. . . . But using the full model set out by Kuhn and adapted by Rorty and Bruffee, a community would also engage

> in abnormal discourse when communicating with people *outside the community* (our emphasis) or when people within the community have differing opinions. I would argue that this state of "abnormal discourse" is actually the usual state of affairs in a discourse community, even if it is unusual in scientific communities (which might also be questioned) [39, p. 149].

Such a model of "autonomous" discourse communities is often represented by a series of circles of varying sizes. The circles are sometimes connected with lines representing some kind of interchange or interrelation (Figure 1), and sometimes the circles are concentric, indicating how certain discourse communities lie within or outside of other communities. Such a model was what we in fact used in our earlier examination of the Project's Phase 4 cross-national survey data. In that examination, as we discussed the effects of language, culture, and sociopolitical conditions on the national and international transfer of scientific and technical information, we constructed a model of the global aerospace science and engineering discourse community in which each local aerospace discourse community was circumscribed (or overlain) by a national community of language, culture, and sociopolitical conditions. These concentric communities were then

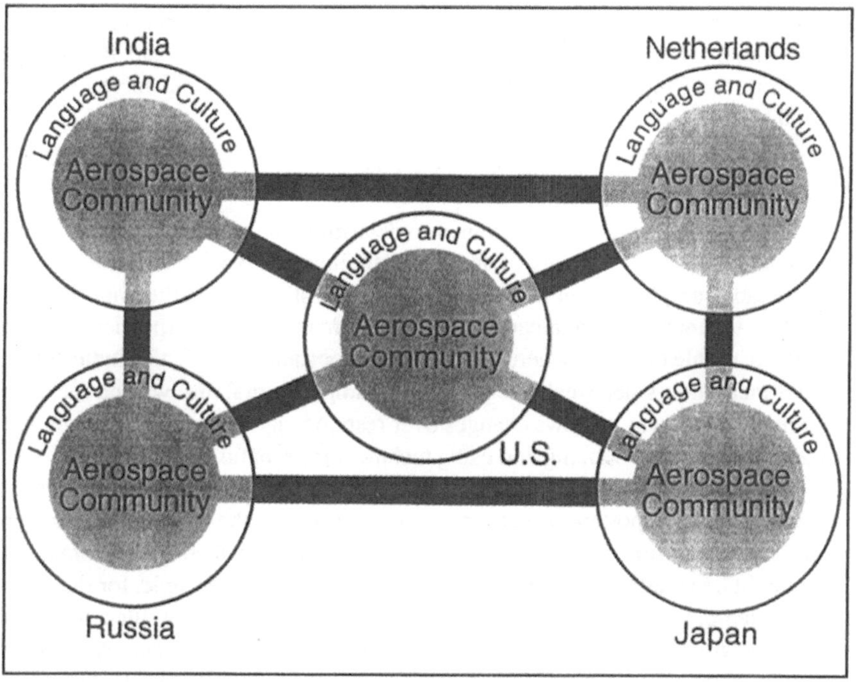

Figure 1.

conceptually interconnected by lines that represented the international transfer of scientific and technical information.

Graphically represented in this way, our model may seem true to the model of discourse communities as adapted by Bruffee and others from Kuhn, which focuses on "normal" discourse as a basis for discourse conventions. However, the interconnecting lines between these national discourse communities would necessarily represent "abnormal discourse"—exchanges of information with others *outside* of the autonomous communities represented by the concentric circles. In retrospect, we do not find that this representation most accurately reflects the model of the international aerospace community of discourse suggested by the Phase 4 survey results.

PHASE 4 SURVEY DATA

Demographics of the Aerospace Professional Survey Samples

Table 1 shows demographic information for the aerospace engineers and scientists responding to survey questions in the Phase 4 cross-national survey of the Project. Responding were 71 aerospace engineers and scientists from India, 94 from Japan, 109 from the Netherlands, 209 from Russia, and 340 from the United States, for a total of five separate national survey samples ($n = 823$).

Importance of Technical Communication on the Job

Parts of the survey were designed to determine the significance of technical communication in the careers of aerospace engineers and scientists. Table 2 first shows each sample's mean response to survey questions about the importance of effective technical communication to the performance of the respondents' duties based on a scale of 1 to 5, where 1 = not very important. Table 2 also indicates the actual mean hours per week spent by each sample in oral and written technical communication, and it shows changes over respondents' careers in the amount of time they spent in producing and using technical information.

The large majority of respondents in each sample rated effective technical communication important or very important. More than their rating of its importance, the actual time the engineers and scientists spent in technical communication might be an effective indicator of its significance. For example, for the entire survey sample taken as a whole, the mean hours spent on written technical communication was nine per week. Compared to the other three survey samples, more Russian and Japanese respondents indicated no change or an actual decrease in the time they spent communicating technical information or working with technical information received from others.

Table 1. Survey Demographics (Percent)

Demographics	India (*N* = 71)	Japan (*N* = 94)	Netherlands (*N* = 109)	Russia (*N* = 209)	U.S. (*N* = 340)
Professional Duties*					
Design/Development	31	31	28	13	6
Administration/Management	6	4	3	2	11
Research	42	33	63	77	82
Service/Maintenance	—	—	—	2	1
Teaching/Academic	20	32	4	1	—
Organizational Affiliation*					
Academic	28	46	—	—	—
Government	65	45	100	100	100
Industrial	—	9	—	—	—
Professional Work Experience					
1-5 years	—	16	37	4	15
6-10 years	17	26	16	22	22
11-20 years	44	32	22	34	28
21-40 years	39	26	25	37	34
41 or more years	—	—	—	3	1
Mean	20	15	12	20	16
Median	20	12	9	17	14
Education*					
Bachelor's degree	7	23	18	51	27
Master's degree	38	35	58	—	46
Doctorate	55	42	19	49	27
Educational Preparation					
Engineer	76	95	74	79	80
Scientists	24	5	25	21	17
Other	—	—	1	—	3
Current Duties					
Engineer	39	100	75	31	69
Scientist	54	—	22	68	27
Other	7	—	3	1	4
Member of a Professional/ Technical Society	85	89	46	22	78
Gender					
Female	6	1	4	15	15
Male	94	99	96	85	85

*Percentages do not total 100.

Table 2. Technical Communication Practices of Aerospace Engineers and Scientists

Communication Factors	India (N = 71) (Mean)	Japan (N = 94) (Mean)	Netherlands (N = 109) (Mean)	Russia (N = 209) (Mean)	U.S. (N = 340) (Mean)
Importance of communicating technical information effectively	4.5	4.7	4.4	4.5	4.5
Hours spent each week communicating technical information in writing	8.0	11.3	9.1	8.3	8.3
Hours spent each week communicating technical information orally	10.0	4.6	6.5	8.7	8.7
Hours spent each week working with written technical information received from others	6.7	6.5	7.4	7.7	7.7
Hours spent each week working with technical information received orally from others	5.1	3.5	4.3	6.3	6.3
Changes Over Time	**Percent**	**Percent**	**Percent**	**Percent**	**Percent**
Compared to 5 years ago, time spent communicating technical information has:					
Increased	83	60	61	30	70
Stayed the same	14	25	35	41	24
Decreased	3	15	4	29	6
As you have advanced professionally, the amount of time spent working with technical information received from others has:					
Increased	75	35	45	38	65
Stayed the same	21	34	50	45	26
Decreased	4	31	5	17	9

Professional Collaborative Writing

A series of questions was formulated to determine the role of collaborative writing in the professional lives of the engineers and scientists responding to the survey. While participation in group writing and the size of the groups varied considerably from sample to sample, Table 3 shows an average of over 90 percent of all respondents who reported that they did at least some of their writing alone, and Tables 4 and 5 indicate that a tenth of the Indian and 7 percent of the Russian respondents, around a quarter of both the Japanese and Dutch, and 15 percent of the U.S. engineers and scientists wrote *only* by themselves. Table 4 also reports the responses of various samples to questions about the relative productivity of collaborative writing practices. Most respondents indicated that collaborative writing was as productive or more so than writing alone.

Professional Use of Technical Information Products

The Phase 4 cross-national survey included a poll of international and U.S. aerospace engineers and scientists on the numbers of each of thirteen specific types of technical information products (10 informal and 3 formal) they had used in the six months prior to the survey. Table 6 gives the mean number of each of the thirteen products used during the six-month period by respondents from the five survey samples.

A few notable differences among the means of product use in the national survey samples:

- Scientists and engineers in India, the Netherlands, and the United States relied on about 11.5 abstracts per month for technical information, while the Japanese used only about three.
- Audiovisual materials, trade literature, and memorandums were seldom used by Russian or Japanese respondents.
- Netherlanders relied on memorandums for technical information much more than did the other survey samples, the Russians hardly ever.
- Indian and U.S. engineers and scientists used less computer programming and documentation.
- Russian respondents relied most heavily on journal articles for their technical information.
- The Japanese preferred conference/meeting papers and in-house technical reports.
- Neither Russian nor Japanese respondents attended many technical presentations.
- Indian engineers and scientists averaged using more technical information products than anyone else, and used nearly as many as the Japanese and Russian respondents combined.

Table 3. Collaborative Writing Practices of Aerospace Engineers and Scientists (Percent)

Collaborative Practices	India (*N* = 71)	Japan (*N* = 94)	Nether-lands (*N* = 109)	Russia (*N* = 209)	U.S. (*N* = 340)
I write alone	93	96	95	79	94
I write with one other person	76	57	65	69	72
I write with a group of two to five people	73	53	50	83	61
I write with a group of more than five people	17	11	10	20	14

*Percentages do not total 100.

Table 4. Influence of Group Participation on Writing Productivity of Aerospace Engineers and Scientists (Percent)

Collaborative Practices	India (*N* = 71)	Japan (*N* = 94)	Nether-lands (*N* = 109)	Russia (*N* = 209)	U.S. (*N* = 340)
A group is more productive than writing alone	52	35	29	44	20
A group is about as productive as writing alone	20	18	20	41	32
A group is less productive than writing alone	18	26	26	8	33
I only write alone	10	21	25	7	15

- Dutch engineers and scientists used more drawings and specifications, technical manuals, computer programs and documentation, and journal articles than any other set of respondents.
- Respondents in the Russian sample did not have a high mean use of any single product.
- Respondents in each of the five survey samples used about the same number of formal information products.

Table 5. Production of Written Technical Communications as a Function Number of Groups and Group Size for Aerospace Engineers and Scientists

Number of Groups	India (N = 71) (%)	Japan (N = 94) (%)	Netherlands (N = 109) (%)	Russia (N = 209) (%)	U.S. (N = 340) (%)
Worked with same group					
Yes	75	48	49	50	47
No	15	31	27	43	38
I only work alone	10	21	24	7	15
Size of Group	**Mean**	**Mean**	**Mean**	**Mean**	**Mean**
Number of people in group					
Mean	4.9	5.1	5.0	3.4	3.2
Median	4.0	3.0	3.0	3.0	3.0
Number of groups					
Mean	1.5	3.1	2.8	2.8	2.8
Median	1.0	3.0	2.0	2.0	3.0
Number of people in each group					
Mean	3.3	3.1	3.5	3.4	3.0
Median	3.0	3.0	3.0	3.0	3.0

Professional Production of Technical Information Products

The respondents were also polled for the number of technical information products that they had produced in the six months prior to the survey. The thirteen types of information products on which they were polled about production were the same as the types of information products on whose use as resources they were polled, and the products were divided into the same categories of informal and formal. The mean number of each type of technical information produced by respondents in the five national survey samples in the six months prior to the survey is listed in Table 7.

Some prominent mean numbers from Table 7 and some comparisons with Table 6:

- Indian respondents produced a high mean number of letters (29.7) as technical information products.
- U.S. engineers and scientists produced a relatively low mean number of letters as technical information products (7.5) compared to the number they

Table 6. Mean Number of Technical Information Products Used in Past Six Months by Aerospace Engineers and Scientists (Mean)

Products	India (*N* = 71)	Japan (*N* = 94)	Netherlands (*N* = 109)	Russia (*N* = 209)	U.S. (*N* = 340)
Informal					
Trade/promotional literature	5.8	4.0	2.1	0.8	5.8
Drawings/specifications	6.8	7.0	7.2	1.7	6.8
Audiovisual materials	3.1	0.5	1.3	0.2	3.1
Letters	23.7	7.6	7.8	2.6	23.7
Memoranda	3.1	1.8	6.4	0.8	3.1
Technical proposals	2.1	2.7	2.1	1.4	2.1
Technical manuals	3.6	4.0	5.8	1.2	3.6
Computer programs/ documentation	2.4	4.9	5.5	4.1	2.4
Technical talks/presentations	3.6	1.9	2.9	1.5	3.6
Abstracts	11.6	2.9	11.1	6.9	11.6
Formal					
Journal articles	14.9	6.6	15.9	14.6	12.3
Conference/meeting papers	8.6	13.1	9.2	3.4	10.0
In-house technical reports	5.0	7.3	5.0	6.2	3.1

Table 7. Mean Number of Technical Information Products Produced in Past Six Months by Aerospace Engineers and Scientists (Mean)

Products	India (*N* = 71)	Japan (*N* = 94)	Netherlands (*N* = 109)	Russia (*N* = 209)	U.S. (*N* = 340)
Informal					
Trade/promotional literature	0.4	3.1	0.2	0.1	0.1
Drawings/specifications	3.0	5.9	1.4	0.9	2.6
Audiovisual materials	2.0	0.1	1.0	0.1	3.9
Letters	29.7	7.9	10.5	1.5	7.5
Memorandums	2.3	1.4	2.7	0.8	9.7
Technical proposals	0.5	3.1	1.2	0.9	0.6
Technical manuals	0.4	1.0	0.3	1.1	0.2
Computer programs/ documentation	0.8	0.7	0.9	1.1	0.7
Technical talks/presentations	3.0	1.4	1.6	0.5	3.2
Abstracts	2.8	1.3	1.0	1.5	1.4
Formal					
Journal articles	0.7	0.9	0.1	0.9	0.7
Conference/meeting papers	2.4	2.2	0.8	0.6	1.1
In-house technical reports	2.5	5.6	1.3	2.4	0.7

used as resources (23.7, the same mean number of letters used by the Indian sample).

- Dutch engineers and scientists produced more letters as technical information products (10.5) than they used as technical information resources (7.8), just as did Indian respondents.
- U.S. engineers and scientists produced more technical memoranda (9.7) than other samples, while the Dutch sample used more of them as information resources (6.4) than other samples.
- Russian respondents had high mean production of computer programs and documentation (1.1).
- The Japanese sample had high mean production of drawings and specifications (5.9), technical proposals (3.1), and in-house technical reports (5.6).
- Russian respondents produced more in-house technical reports (2.4) than journal articles and conference papers.
- Dutch respondents produced few (0.1) journal articles each, while reading more than anyone else (15.9).
- Japanese and Russian respondents each produced a mean high number of journal articles (0.9); the Japanese used a mean of only 6.6 journal articles to the Russians' 14.6.
- The Indian and Japanese samples produced more conference papers (2.2, 2.4, respectively) than the others; the Japanese read more conference papers than any other sample (13.1).

Professional Sources of Scientific and Technical Information

Phase 4 cross-national survey respondents were asked a series of questions concerning the sources of information that they used in their duties as aerospace engineers and scientists in their respective countries. Table 8 lists the percentages of respondents using five different sources of information that were applied to problem solving in their work. Their frequency of use of libraries as a source of information is shown in Table 9. Tables 10 and 11 illustrate the respondents' use and evaluation of the importance of technical reports from their own and the national aerospace organizations of other countries.

Some salient facts gleaned from Tables 8 and 9:

- Generally, respondents reported a high dependence (almost 100%) on their own information resources of whatever kind (Table 8).
- Equally high percentages of all national samples relied heavily on oral communication with peers (Table 8).
- Only half of the Russian scientists and engineers reported using a personal store of scientific and technical information (Table 8).

Table 8. Information Sources Used by Aerospace Engineers and Scientists in Problem Solving (Percent)

Source	India (*N* = 71)	Japan (*N* = 94)	Netherlands (*N* = 109)	Russia (*N* = 209)	U.S. (*N* = 340)
Personal store of technical information	96	97	98	51	99
Spoke with a coworker or people inside my organization	93	94	98	96	99
Spoke with a colleague outside my organization	75	81	79	36	93
Used literature resources found in my organization's library	94	72	95	85	91
Spoke with a librarian or technical information specialist	55	50	74	59	80

Table 9. Use of a Library in the Past Six Months by Aerospace Engineers and Scientists (Percent)

Visits	India (*N* = 71)	Japan (*N* = 94)	Netherlands (*N* = 109)	Russia (*N* = 209)	U.S. (*N* = 340)
0 times	1	12	5	5	11
1-5 times	4	16	20	32	44
6-10 times	9	29	28	35	22
11-25 times	35	20	35	20	15
26-50 times	38	16	6	7	7
51 or more times	21	7	6	1	1
Mean	43.0	20.9	18.5	11.3	9.2
Median	27.5	10.0	10.0	0.0	4.0

Table 10. Use of Foreign and Domestically Produced Technical Reports by Aerospace Engineers and Scientists (Percent)

Country/Organization	India (N = 71)	Japan (N = 94)	Netherlands (N = 109)	U.S. (N = 340)
NATO AGARD reports	69	60	71	82
British ARC and RAE reports	75	48	50	54
ESA reports	35	25	44	6
Indian NAL reports	79	3	7	6
French ONERA reports	44	39	43	41
German DFVLR, DLR, and MBB reports	58	53	69	36
Japanese NAL reports	18	87	11	12
Russian TsAGI reports	3	2	1	8
Dutch NLR reports	31	23	96	20
U.S. NASA reports	96	89	82	97

Table 11. Importance of Foreign and Domestically Produced Technical Reports to Aerospace Engineers and Scientists (Rating Mean)*

Country/Organization	India (N = 71)	Japan (N = 94)	Netherlands (N = 109)	U.S. (N = 340)
NATO AGARD reports	4.30	3.67	3.18	3.65
British ARC and RAE reports	4.16	3.12	2.87	3.22
ESA reports	3.77	2.78	2.35	1.52
Indian NAL reports	3.97	2.02	1.46	1.51
French ONERA reports	3.25	2.97	2.36	2.48
German DFVLR, DLR, and MBB reports	3.50	3.15	3.22	2.40
Japanese NAL reports	2.63	3.94	1.57	1.75
Russian TsAGI reports	2.15	2.23	1.31	1.81
Dutch NLR reports	3.03	2.65	4.32	1.95
U.S. NASA reports	4.47	4.46	3.69	4.37

*A 5-point scale was used to measure importance with 1 being the lowest possible importance and 5 being the highest possible importance. Hence, the higher the average (mean) the greater the importance of the report series.

- Very few Russian scientists or engineers spoke with colleagues outside their organizations (Table 8).
- Except for Japan (72%), very high percentages of all survey respondents used literature from their own organizational libraries (Table 8).
- Only in the Netherlands and the United States did more than half of any national sample seek information by speaking with a librarian (Table 8).
- Indian scientists and engineers used their libraries more times on average (43.0) in the six months prior to the survey, twice as much as any other national sample; about 35 percent of the sample used Indian libraries eleven to twenty-five times, 28 percent used them twenty-six to fifty times, and 21 percent used them fifty-one or more times—no other sample came close (Table 9).
- Respondents in the U.S. national sample used library resources less than any other (a mean of 9.2), most of them only one to five times. Mean library use by the other three samples was distributed fairly evenly between these two extremes (Table 9).

Some interesting patterns emerge from the responses of the national samples that ranked by percentage (Table 10) and rated the mean importance of (Table 11) their use of technical reports from their own and other aerospace organizations:

- There were no reports from the Russian sample on their use of any aerospace organization reports.
- Very few (a mean of less than 4%) of the respondents from the other national samples reported using Russian Central Aero-Hydrodynamics Institute (TsAGI) reports.
- Reports from NASA and NATO AGARD got the highest percentage of use across the board and also received the highest marks for their importance to the engineers and scientists using them.
- Reports from the German aerospace organization ranked and rated high, too, followed in rough order by the British, the French, the Dutch, and reports from the European Space Agency (ESA).
- Percentages of respondents using particular organizational reports roughly paralleled the ranking of their importance.

Professional Recommendations for an Undergraduate Technical Communication Course

The engineers and scientists responding to the Phase 4 cross-national survey were asked to recommend communication principles, language mechanics, and technical information product skills to be taught in an ideal, realizable undergraduate technical communication course for aerospace science and engineering students. Their recommendations are listed in Table 12 by the percentage of

Table 12. Principles, Mechanics, and On-the-Job Skills Recommended for Inclusion in an Undergraduate Technical Communications Course by Aerospace Engineers and Scientists (Percent)

	India ($N = 71$)	Japan ($N = 94$)	Netherlands ($N = 109$)	Russia ($N = 209$)	U.S. ($N = 340$)
Principles					
Defining the purpose of the communication	87	65	89	63	91
Assessing the needs of the reader	59	44	83	57	87
Preparing/presenting information in an organized manner	90	75	83	64	97
Developing paragraphs (introductions, transitions, conclusions)	76	90	89	77	87
Writing grammatically correct sentences	52	85	57	—	72
Notetaking and quoting	35	51	41	69	44
Editing and revising	72	51	41	59	87
Mechanics					
Abbreviations	56	66	47	70	55
Acronyms	39	64	39	43	52
Capitalization	35	50	31	39	54
Numbers	35	51	33	43	48
Punctuation	51	53	54	35	74
References	75	68	63	76	80
Spelling	47	44	58	37	55
Symbols	58	66	53	61	64
On-the-Job Skills					
Abstracts	76	48	82	84	85
Letters	38	27	51	75	61
Memoranda	28	25	66	53	60
Technical instructions	56	59	63	64	62
Journal articles	69	48	49	92	64
Conference/meeting papers	66	78	54	79	67
Literature reviews	62	21	39	77	50
Technical manuals	61	56	60	54	43
Newsletter/newspaper articles	32	9	17	62	15
Oral (technical) presentations	78	72	84	80	92
Technical specifications	58	60	56	57	45
Technical reports	83	70	86	81	81
Use of information sources	59	60	72	73	72

respondents in each sample who endorsed a particular principle, mechanic, or skill.

Technical writing principles recommended by many members of all the survey samples for inclusion in the ideal course included:

- defining the purpose of the communication,
- organizing the information, and
- paragraph development.

Many respondents also recommended teaching editing and revision and grammatical accuracy. Overall, the most recommended course component in document mechanics was the use of references. Writing abstracts and giving oral technical presentations were the on-the-job skills most recommended to be taught as part of an undergraduate technical communication course.

Findings and Limitations

Table 1 indicates that the typical professional respondent to the survey was male and was likely to be an engineer rather than a scientist. The gender ratio varied in a somewhat predictable way—that is, the number of women in a particular survey sample generally reflected our cultural assumptions about the opportunities for women in that country to enroll in higher education and advance in technical careers; thus it surprised us that the Netherlands sample did not include more women. Most respondents were employed in government organizations; except for Russians, many had or would attain advanced degrees—the Russian bachelor's degree in its requirements resembles a U.S. master's degree, with more industry experience [41].

Only 21 percent of the Russian respondents were educated as scientists, while 68 percent reported having current duties as scientists (Table 1). From a cultural perspective, this may reflect the R&D mission of Russian government organizations with which they are affiliated, or it may imply a different way of distinguishing between "engineer" and "scientist" in Russian aerospace organizations, or it may highlight an emphasis on engineering rather than science in Russian academies. Similarly, 54 percent of the Indian respondents reported duties as scientists (Table 1), although only 24 percent were educated as scientists. This could be related to the higher mean and median years of work experience of the Indian sample.

All the respondents were in accord over the importance of communicating technical information (Table 2). There were, however, significant differences in how they received and imparted information (Tables 2, 6, 7, and 8). Once again focusing on the culture to explain these differences, the low Russian figure for using letters as a source of technical information (Table 6) might relate to, for example, the efficiency of their postal infrastructure. Information use and production by different survey samples can also differ in ways not represented by the survey data. For instance, due to a lack of infrastructure, Indian database users depend largely on hard copy versions of those databases [42].

It would be wrong to turn survey results like those shown in Table 7 (production of technical information products) into an international competition. Although comparisons are bound to be made, for example, between the gross output of one national sample and any or all of the others, comparing the production of written technical information by these national samples would seem to require the utmost discrimination among the demographic, linguistic, and cultural characteristics of the survey samples.

For example, working in groups rather than working alone will yield a different document output, and most of the respondents agreed that collaborative writing was as productive or more productive than writing alone (Table 4). But the Japanese scientists and engineers—who, from the literature on communication in Japanese organizations [11, 43], might be expected to collaborate more and more often—collaborated less when producing written technical communication than did the scientists and engineers in the other survey samples (Tables 3 and 5). What should we make of that? We believe there are simply too many variables—cultural and organizational—to use the mean number of written technical documents produced by an individual professional or group of a given type within a given period as a yardstick for the output of aerospace scientific and technical information by a particular nation's aerospace industry.

Other examples of misreading a communication practice through a cultural lens might involve how one sees the Japanese sample's low means for both giving and receiving oral technical communication made by Haas and Funk, but it could be a result of linguistic differences: the Japanese respondents may not have the same definition of oral technical communication as the Western observers. The higher Indian mean for communicating orally with others (Tables 2 and 8) may reflect a cultural propensity for oral communication [44, 45], or it may be related only to the Indian sample's high mean (20) years of professional experience (Table 1). That is, it might be expected that the senior engineers and scientists would relay instructions orally to subordinates. The fewer Russian scientists and engineers (about half) who reported using a personal store of information (Table 6) may relate to the high level of control historically exerted on information acquisition and dissemination in the former USSR [46], or it may indicate only that the Russian aerospace professionals depend more on information from their organizational sources, such as libraries (Table 8).

All the surveyed professionals depended a great deal on oral communication with their peers as a source of scientific and technical information (Tables 2 and 8). But Table 8 also shows that the overall majority of the aerospace professionals relied primarily on their own personal collections of technical information, implying a certain amount of independence in gathering that information. For each survey sample, English-language aerospace agency reports generally ranked higher in importance as sources of technical information (Table 11).

In recommending a curriculum for a technical communication course (Table 12), the Russian sample diverged the most from the recommendations of the other four national samples, although there are singularities throughout Table 12. Japanese respondents, for example, might be expected—based on the literature [47]—to devalue the importance of assessing the readers' needs, and they did. Another example: Russian aerospace professionals might not place a high value on the teaching of grammar because it is not part of the curriculum in Russian secondary and higher education.

Significance

Overall, the responses of the aerospace professionals to each of the survey questions did not seem to us to vary significantly from sample to sample in a way that could clearly be attributed primarily to linguistic, cultural, and sociopolitical differences. As we have pointed out above, there were certain patterns of response by different survey samples from which inferences could be drawn about the effects of language, culture, and sociopolitical conditions on their communication practices; but just as often, our intuition—indeed, our bias—from communication and composition research that a particular country's native language or culture would affect that country's (predominantly English-language, aerospace) technical communication practices in a certain way was not verified by the data. Taken as a whole, the Phase 4 survey results indicated that, worldwide, given the same technological opportunities, aerospace professionals are similar in their communication practices. It was this recalcitrance of the data to fit our assumptions that tipped us toward the following reconsideration of the effects of language, culture, and sociopolitical conditions on international professional communication in general.

CONCLUDING REMARKS

One possibility is that the survey questions we asked, and perhaps the survey methodology itself, simply are not sensitive enough to detect differences among the cultures. The differences are not quite big enough, the questions and methodology too blunt. Testing for this explanation's appropriateness obviously must await further research. Here we choose to explore the other possibility. What if the language (English) and culture (aeronautical engineering and science) of the professional communities being studied here in fact override most differences caused by native language and culture?

If, on a theoretical plane, we are not content with the model of the worldwide aerospace discourse community represented by interconnected, concentric circles as in Figure 1, and if the Phase 4 data reinforce our discontent, then what? Our first consideration would be to invert the circumscription of the local aerospace communities by their national languages and cultures, so that the communities of aerospace engineers and scientists now outlie their respective languages and cultures, which then lie at the cores of the aerospace communities (Figure 2). This arrangement makes more sense. The interconnections are more direct: communication is direct among the different local aerospace communities without being filtered through a layer of language and culture, as it is in Figure 1. The communities of aerospace professionals represented in Figure 2 could share a world view, values, and a way of legitimating knowledge—and in that sense, a *language*. In Figure 2 we have created a successful model of an international scientific community, at least.

But something about this Figure 2 model still does not work as a representation of international communication between the members of this scientific community, because it unnecessarily fragments the global aerospace community. For example, these autonomous communities of aerospace scientists and engineers interact only in a realm exclusive of their native languages and cultures. This representation does not yield a model that encompasses both the "normal" and "abnormal discourse" that occurs between aerospace professionals who are members of the same and of different local communities. International communications in Figure 2, still represented here by interconnecting lines, are

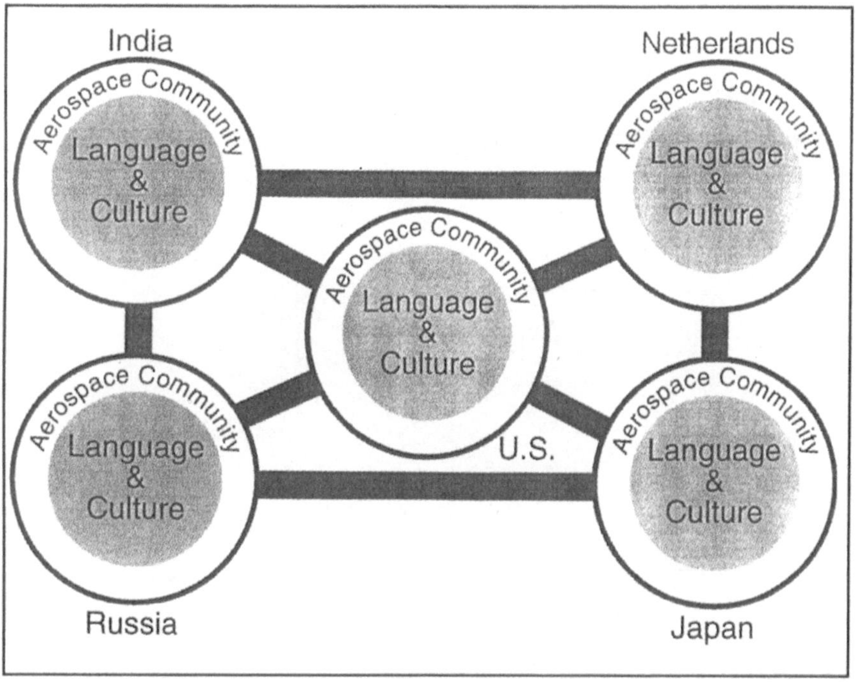

Figure 2.

modeled, also as in Figure 1, as "abnormal discourse," exchanges of information with others *outside* of the autonomous communities represented by the concentric circles.

To account successfully for the entire global aerospace community's necessary international communications, we can model that community itself as a single domain that circumscribes a lesser realm of language and culture (Figure 3). The model is now one of a complete global community, totally encompassing both "normal" and "abnormal" discourse, but something is still lacking. How do we correct the model for the effect upon the global aerospace community's international communications of the many different languages and cultures represented by the circle within? The unifying factor that brings this model into line is the use of the *English language* in international communications by the engineers and

Figure 3.

scientists in the global aerospace community. Thus, the "discourse community" model of international professional communication works when English is the common tongue.

Still, something bothers us about the model of community exhibited in Figure 3 as a metaphor for international communication. Maybe it is just too simple. As Longo [48] explains, what gives an abstract grouping of individuals, a discourse community, existence is the shared purposes, values, and assumptions that give rise to a shared discourse: a person becomes a member of the community upon demonstrating proficiency in the shared discourse. Individuals *choose* to join the discourse community of aerospace scientists and engineers—or of certified public accountants, computer programmers, salespeople, university professors, or rhetoricians [48, p. 397]. But how do we account, in our model of international communication, for concrete—not abstract—groupings of people, such as ethnic and linguistic groups [48, p. 400]? These groupings are based not on choice but on *chance*. Circumscribing "language and culture," a chance grouping, with "the discourse community," a grouping of choice, is not entirely adequate; the subsequent model is not sufficiently robust to accommodate the global aspects of international communication.

Longo [48] suggests that we need a concept of communication that is both flexible and local in time and space. In such a model, any consensus achieved by a group would be local in time and space, a "temporary contract" [49] that would enable people from different communities to enter into and dissolve the contract [48, p. 399]. This "temporary contract" is very much the manner in which people, including aerospace engineers and scientists and professional communicators of all stripes, currently communicate, for example, over the Internet in its various incarnations, including but not limited to e-mail. Longo cites Killingsworth's [50] discussion of local and global communities as dialogic counterparts within the idea of community. "Breaking up the idea of community in sub-communities that are necessarily in conflict (as well as in consensus) is a first step towards reconceiving communication as a fragmented, postmodern practice." But, Longo adds, Killingsworth's example of local and global communities still implies clear boundaries around groups of people in these communities. "A postmodern model of communication will include boundaries that are not stable . . . and groups that come together and dissolve, sometimes within a short time" [48, p. 399]. In fact, Longo describes what looks to be a perfect model for electronic networking.

Longo expands on what a "postmodern" model of communication would be:

> We can describe various dimensions of communication situations as continua along which these situations can be located. For example, we could set out these dimensions: choice/chance, time/space, abstract/concrete, affinity/proximity. We could then describe a disciplinary communication grouping as determined more by choice than by chance, more constrained by time than by place, more abstract than concrete, more determined by affinity than proximity. We could possibly describe an ethnic grouping as determined

> more by chance than choice, more constrained by place than time, more concrete than abstract, more determined by proximity than by affinity. . . . By using a more fluid approach to modeling communication, we acknowledge the local situatedness of communications and allow for flexibility in conceptualizing them. We break up the either/or, clearly bounded approach to modeling community and better accommodate the ambiguity we intuitively understand as an important part of communication [48, pp. 399-400].

We intuited such an ambiguity when the accepted "community" model of communications could not suitably accommodate what our data show taking place in international communication within the global aerospace community. It is our current intuition that the accepted "community" model may not be robust enough to accommodate the contemporary (synonymous with international) professional communication of any discipline. Our earlier (Figure 1) model is clearly limited by its domination by boundaries; Figures 2 and 3 in lesser ways display the same limitation. The disadvantage of using such models is that we then shape our research as though boundaries are similarly primary in the reality we seek to describe through our data. The possibility we advance here, that the really significant features are not those which are boundaried (each nation's language and culture) but those which transcend boundaries (the language and culture of aeronautical engineering and science), would allow exploration of the kinds of models Longo invites, arrays of continua along which various communication situations may be located.

As new media and genres (such as those that have proliferated as a result of electronic networking) continue to reshape our world, professional communicators must expand their understanding to encompass newer, more flexible models of communication that allow room for data which may contest our field's current paradigms. In those new models, factors such as those that Longo discusses above may well turn out to be more important than country of origin or language first acquired.

REFERENCES

1. T. O. Pinelli, J. M. Kennedy, and R. O. Barclay, The NASA/DoD Aerospace Knowledge Diffusion Research Project, *Government Information Quarterly, 8*:2, pp. 219-233, 1991.
2. J. R. Kohl, R. O. Barclay, T. E. Pinelli, M. L. Keene, and J. M. Kennedy, The Impact of Language and Culture on Technical Communication in Japan, *Technical Communication, 40*:1, pp. 62-73, February 1993.
3. K. A. Bruffee, *Collaborative Learning—Higher Education, Interdependence, and the Authority of Knowledge,* Johns Hopkins University Press, Baltimore, Maryland, 1993.
4. T. E. Pinelli, R. O. Barclay, M. L. Keene, J. M. Kennedy, and L. F. Hecht, From Student to Entry-Level Professional—Examining the Role of Language and Written Communications in the Reacculturation of Aerospace Engineering Students, *Technical Communication, 43,* pp. 492-503, August 1995.

5. J. R. Webb and M. L. Keene, The Production and Use of Information Among Aerospace Engineering Professionals and Students—A Cross-National Analysis, Chapter 14 in *Knowledge Diffusion in the U.S. Aerospace Industry,* T. E. Pinelli (ed.), Ablex Publishing, Norwood, New Jersey, 1997.
6. T. M. Madden, *Cultural Context and Informational Assumptions as Sources of Variation in Social Reasoning in India,* Ph.D. dissertation, University of California, Berkeley, 1992.
7. J. Bakshi, India's Geopolitical Importance in the United Nations, *India Quarterly, L*:4, pp. 93-98, October-December 1994.
8. DOC (U.S. Department of Commerce), Chapter 20 Aerospace, *U.S. Industrial Outlook 1994,* U.S. Government Printing Office, Washington, D.C., pp. 201-211, January 1994.
9. J. Fischer, *Personal Communication to V. Bolish, Director,* Faculty Center, California State University, San Marcos, Congressional Research Center, Washington, D.C., December 13, 1995.
10. J. R. Webb, *Writing Practices of ESL Aerospace Engineering Students: An Evaluation of the Phase 3 Survey Results of the NASA/DoD Aerospace Knowledge Diffusion Research Project,* unpublished manuscript, University of Tennessee, Knoxville, 1995.
11. D. W. Stevenson, Audiences Analysis Across Cultures, *Journal of Technical Writing and Communication, 13*:4, pp. 319-330, 1983.
12. R. W. Bailey and M. Gorlach (eds.), *English as a World Language,* University Press, Cambridge, 1993.
13. J. Chesire (ed.), *English Around the World: Sociolinguistic Perspectives,* University Press, Cambridge, 1991.
14. S. Greenbaum (ed.), *The English Language Today,* Pergammon, Oxford, 1985.
15. B. B. Kachur (ed.), *The Other Tongue: English Across Cultures* (2nd Edition), University of Illinois Press, Urbana, 1992.
16. M. Berns, English in the European Union, *English Today 43, 11*:3, pp. 3-11, July 1995.
17. B. Edens, Master One's First Language: The Implications of Writing in International English, *Proceedings of the 1992 International Professional Communication Conference, Santa Fe, New Mexico,* pp. 538-542, September 29-October 3, 1992.
18. S. Ridder, English in Dutch, *English Today, 44,* pp. 43-50, October 1995.
19. J. T. Dennett, World Class Status Does Not Ensure World Class Usage, *IEEE Transactions on Professional Communication, 35*:1, p. 13, March 1992.
20. NCTE (National Council of Teachers of English), What Makes Writing 'Good'? International Panel Disagrees, *The Council Chronicle, 1,* pp. 1-3, September 1991.
21. M. Subbiah, Adding a New Dimension to the Teaching of Audience Analysis: Cultural Awareness, *IEEE Transactions on Professional Communication, 35*:1, pp. 14-18, March 1992.
22. C. Z. Zee, Multiculturalism in the Professional Communication Curriculum, *IEEE Transactions on Professional Communication, 35*:3, p. 138, September 1992.
23. R. Schechner, TDR Comment, *The Drama Review, 36*:1, p. 7, Spring 1992.
24. A. Bloom, *The Closing of the American Mind,* Simon and Schuster, New York, 1987.
25. D. S. Bosley, Cross-Cultural Collaboration: Whose Culture Is It, Anyway? *Technical Communication Quarterly, 2,* pp. 51-62, 1992.

26. S. Thiederman, *Profiting in America's Multicultural Marketplace,* Macmillan, New York, 1991.
27. J. Bisong, Language Choice and Cultural Imperialism: A Nigerian Perspective, *English Language Teacher's Journal 49*:2, pp. 122-132, 1995.
28. R. Phillipson, *Linguistic Imperialism,* Oxford University Press, Oxford, 1992.
29. M. Dakich, Oral Language from Different Cultures: Surveying and Identifying Problems Experienced by Technical Personnel, *Proceedings of the 1992 International Professional Communication Conference, Santa Fe, New Mexico,* pp. 562-565, September 29-October 3, 1992.
30. B. McDaniel, The Role of Contrastive Rhetoric in Teaching Professional Communication in English as a Second or Foreign Language, *IEEE Transactions in Professional Communication, 37*:1, pp. 29-33, March 1994.
31. H. Yamada, *American and Japanese Business Discourse: A Comparison of Interactional Style,* R. O. Freedle (ed.), Ablex, Norwood, New Jersey, 1992.
32. B. McDaniel, Translating Thoughts into English Structure: The Challenge for Nonnative Speakers of English, *Proceedings of the 1992 International Professional Communication Conference, Santa Fe, New Mexico,* pp. 566-572, September 29-October 3, 1992.
33. C. Boiarsky, Strategies for Successful International Communication: Using Cultural Conventions to Inform Your Documents, *Proceedings of the 1992 International Professional Communication Conference, Santa Fe, New Mexico,* pp. 553-559, September 29-October 3, 1992.
34. J. M. Uljin, J. Hoppenbrouwers, and G. Mulder, Writing for a Client in International Business and Technology: Does Culture Affect His/Her Expectations? Some Evidence from Reading Industrial Research Proposals and Technical Manuals, *Proceedings of the 1992 International Professional Communication Conference, Santa Fe, New Mexico,* pp. 573-578, September 29-October 3, 1992.
35. C. P. Campbell and P. Bernick (eds.), 'Good English,' and International Readers, *Proceedings of the 1993 International Professional Communication Conference,* Philadelphia, Pennsylvania, pp. 38-42, October 5-8, 1993.
36. K. A. Bruffee, Social Construction, Language, and the Authority of Knowledge, *College English, 48,* pp. 773-790, 1986.
37. K. A. Bruffee, Collaborative Learning and the 'Conversation of Mankind,' *College English, 46,* pp. 635-652, 1984.
38. T. S. Kuhn, *The Structure of Scientific Revolutions* (2nd Edition), University of Chicago Press, Chicago, 1970.
39. B. C. Longo, Teaching Technical Communication with the Community Model: Some Questionable Side Effects, *Proceedings of the 1995 International Professional Communication Conference, Savannah, Georgia,* pp. 148-151, September 27-29, 1995.
40. R. Rorty, *Philosophy and the Mirror of Nature,* Princeton University Press, Princeton, New Jersey, 1979.
41. A. Shapero, Life Styles of Engineering, *Space/Aeronautics, 51*:3, pp. 58-65, March 1969.
42. A. Lahiri, *Initiatives to Facilitate Access to S&T Information in India,* paper presented at the International Federation of Library Associations and Institutions (IFLA) 58th General Conference, New Delhi, India, pp. 181-202, August 30-September 3, 1992.

43. C. Haas and J. L. Funk, 'Shared Information': Some Observations of Communication in Japanese Technological Settings, *Technical Communication, 36*:4, pp. 362-367, 1987.
44. S. Rao, Contemporary Orality: A New Theory for Understanding Speech, paper presented at the 77th Annual Meeting of the Speech Communication Association, *Florida Communication Journal, 20*:1, p. 10, 1991.
45. S. Rao, *Contemporary Orality: The Understanding of a Consciousness in Urban India (Urban Culture)*, Ph.D. Dissertation, University of Massachusetts, Amherst, 1993.
46. R. O. Barclay, T. E. Pinelli, M. Flammia, and J. M. Kennedy, *The Impact of Political Control on Technical Communications: A Comparative Study of Russian and U.S. Aerospace Engineers and Scientists*, paper presented at the 32nd Aerospace Meeting and Exhibit of the American Institute of Aeronautics and Astronautics, Reno, Nevada, January 11, 1994.
47. J. T. Dennett, Not to Say is Better Than to Say: How Rhetorical Structure Reflects Cultural Context in Japanese-English Technical Writing, *IEEE Transactions in Professional Communication, 31*:3, pp. 116-119, 1988.
48. B. C. Longo, Choice or Chance: Questioning Dimensions Within the Idea of Community, *Journal of Technical Writing and Communication, 25*:4, pp. 393-400, 1995.
49. J. F. Lyotard, *The Postmodern Explained*, University of Minnesota Press, Minneapolis, 1992.
50. M. J. Killingsworth, Discourse Communities—Local and Global, *Rhetoric Review, 11*:1, pp. 110-122, Fall 1992.

CHAPTER 5

Redefining the Professional in International Professional Communication

TIMOTHY S. BOSWOOD

In the movie *Mrs. Doubtfire,* comedian Robin Williams is divorced by his thirty-something wife, a very successful interior designer, who is then commissioned by an attractive ex-boyfriend to work on a new development project. He asks her if the working relationship will be a problem for her. She answers, flustered and blushing, and avoiding his eyes, "Of course not. I'm a professional." The line is funny, since she is obviously both bothered and flattered by his attentions. Where does the humor lie?

Like much comedy, the success of the line rests on contradiction, human weakness, and failure. The humor derives from the ideology of professionalism, which, while promoting the individual as organizational heroine, values also the subordination of personal emotions and desires to the needs of the organization and to the practices of the professional group. The way the woman delivers the line contradicts its meaning, and so we laugh. Though she claims the mantle of professionalism she demonstrates her unworthiness to bear it, and we laugh. In attempting to establish her professional identity she reveals feet of clay, and we laugh.

My aim in this chapter is to redefine the nature of professionalism as it applies to professional communication, in the hope that it will lend direction to the teaching of professional communication in an international context. In doing this I will argue that professional communication training is essentially preparation for the management of identities in interdiscourse contacts and further that it is the business of the trainer and trainee to develop critical awareness of the identity or identities that the training process espouses.

I will attempt this clarification by developing nine propositions about professionalism. These propositions, and the discussion that leads to each, are based on a social analytic approach that places the professional in a social context, and looks to the techniques of discourse analysis and the ethnography of communication for an understanding of this context. This is clearly becoming a powerful paradigm in the analysis of professional communication [1-4].

In their survey of research directions with a social orientation, Thralls and Blyler recognize the fruitfulness of ideologic analyses of professional discourse but doubt that their pedagogic applications can go much beyond "enlightening students about some issues of gender, race, and class, and the tendency for certain voices to be marginalized in discourse situations" [5, p. 32]. They doubt further application of an ideologic approach, since it is likely to bring the practitioner into non-productive conflict with empowered interest in both industry and the academy. This appears to me to be a particularly Western position that assumes the individual to be in essential conflict with overarching social groups such as governments and corporations. This assumption is not shared globally, and is questionable especially in my own teaching context in East Asia. A further aim of this chapter, then, is to indicate ways in which this confrontation can be turned to constructive ends, by reconciliation and synthesis, through a redefinition of the concept of professionalism in an international context. This chapter, therefore, also draws on research in intercultural communication for guidance in developing educational programs for both native and non-native speakers of the languages in which the target professional discourse is realized.

THE PROFESSIONAL: ASPIRING TO A STEREOTYPE

When we look at professionalism we are teasing out the elements of a stereotype. It is a stereotype with particularly Western characteristics, and roots, it is fair to guess, in the ideology of utilitarianism [4]. In this respect, the professional is a caricature, represented and reinforced internationally by public communications, partly by corporate messages, but also, as in the example that opens this chapter, by the widely-exported United States entertainment media. Like any stereotype, it embodies an ideology. By using the phrase *professional communication,* we implicitly subscribe to this stereotype, to the values contained in the ideology, and we invite our students to aspire to it.

In most cases, this ideology is taught covertly. The notion of professionalism receives scant explicit analysis in standard undergraduate textbooks on communication. One example, from an otherwise useful text, deals with the notion of professionalism in two-and-a-half pages, under the headings of office politics, networking, and personal appearance [6, pp. 241-243]. Many leading textbooks have no index reference for professionalism [7-10]. Even some which appeal to professionalism in their titles present no analysis of the concept and rarely use the term in the body of their texts [11, 12]. A rare exception to this is Huckin and Olsen whose opening chapter analyzes technical communication as social interaction and touches on some of the issues raised in this chapter [13]. Significantly, these authors, unlike those cited above, are writing for non-native speakers of English. It is precisely at such points of cultural contact, in the international arena, that notions such as professionalism come into question. Generally, *professionalism*, as a component of a dominant domestic ideology, is taken for granted and used almost interchangeably with *quality*.

The focus of these instructional texts is on developing the skills to be wielded by the person, not on the person who develops through the wielding of skills. I do not know of any way, however, to divorce the communication skills that we teach from the values of the identity from which they derive. It is essential, therefore, that we question this covert stereotype and consider carefully its relevance to our students and the workplace they are entering. For beyond the comfortable certainties of academic texts lie the demands of the newly globalized workplace, where disparate communities and organizations apply diverse standards, and where rewards are given for, and sanctions are enforced against, behaviors which may or may not correspond to a Western, academic stereotype of professionalism.

TOWARD A NON-ELITIST SENSE OF PROFESSIONALISM

Professions, Practitioners, and Elites

As a starting point in this inquiry we can ask: What is a *profession*?

The traditional concept of a profession is that of a self-governing occupation. Originally, this referred to the three learned disciplines: the church, the law, and medicine. More recently, however, *profession* has come to refer to any occupation based on the mastery of a discipline which is self-governing in the sense that it has an association, i.e., an organization which issues accreditation, and sets and applies *standards of practice*. Thus, professionals are often called *practitioners* (e.g., legal practitioners, medical practitioners). There are professional associations for accountants, architects, and engineers, but also for public relations officers, human resource managers, and marketing managers. Swales concurs that these occupations, and others like them, should be included within the ranks of the professions: "Professions will be understood in a non-elitist sense which includes service and other occupations . . . as well as the more prototypical categories of medicine, law, etc." [14, p. 103]. Though this process may be deplored as voiding the special role of the established professions in society [15, p. 96], it is a *fait accompli*.

Professionalization is the coming of age of trades. It is the formation of a more or less elite club. Is communication a profession in this sense? Clearly not, in the sense that there is no specific professional association for graduates in professional communication—though their teachers may belong to several! So what can *professional communication* mean? To further the inquiry, we can distinguish three modes in which professional people relate to their own communication.

Three Modes of Professional Communication

Doctors and lawyers, referred to above as the prototypical professionals, are *professionals who communicate.* We may add engineers and many of the scientific and technical professions to this list. In general, members see their profession as involving communication, but usually as peripheral to their main professional allegiance. Communication ability is a part, but not usually the central part, of their professional identity.

But another group of professionals sees communication as central to the performance of their role—it is a core element in the professional identity. These are members of the so-called *communication professions,* including the media, public relations, advertising, human resources management—sometimes (in Britain at least) referred to as *the chattering classes.* By the mid-eighties, it was clear that the demand for competent employees in these sectors was increasing (e.g., see [16] for a survey of the Australian scene). A similar increase in Hong Kong has also been predicted [17]. The situation is likely to be similar in the other fast-developing economies of the Pacific Rim.

A third, and new, dimension is also suggested by developments in the Western industrialized countries where the crucial role played by communication in organizations is increasingly appreciated. Research, particularly in Australia, has indicated the emergence of a new area of expertise, communications management, and a job to go with it, *the communications manager.* The extracts below summarize some aspects of the work of such a professional.

> Communication management is emerging as an important managerial role in private and public sector organizations. The role of the communication manager has tended to cut across traditional boundaries in areas such as public relations, human resource development, media management, issues management, and marketing [18, p. 113].

> The work areas or departments in which Australian communication managers were employed included public relations, communications, marketing, corporate relations, public affairs, personnel/human resources, public information, video, internal communications, and community relations [19, p. 18].

> Communication management is defined as broadly concerned with managing and administering communication resources and processes to facilitate communication within organizations and between organizations and their communities [20, p. 81].

> Communication managers . . . are required to evaluate organizational communication processes and to develop intervention strategies to improve them. Their broad goal concerns the development of an organizational communication climate which supports organizational development [18, pp. 113-114].

In the emerging cross-disciplinary role of the communications manager, the professional not only has communication practice as a central element in the professional identity, but is also concerned with managing the communication practices of others, possibly of a whole organization.

To summarize, then, professionals may engage with communication issues in three modes: as professionals who communicate, as professional communicators, and as communications managers. It is with the latter two that I am concerned here in the discussion of professionalism.[1]

A Preliminary Definition of Professionalism

When we talk about a person communicating professionally, we can make a distinction of meaning as follows, between:

1. communicating as a professional; and
2. communicating to a professional standard.

In the first meaning, we are asserting that a person, through communication style, is playing a professional role, or claiming a professional identity, e.g., communicating in the role of a communication manager, of a marketing manager, as opposed to presenting a private, individual persona [21]. Although much of what we may call professional communication is of this kind, no doubt there are also very interesting occasions where the strategy of throwing aside the professional persona and communicating "as an individual" is employed to good effect. An interesting area of research would be to inquire into the ways in which this strategy is used in various societies, particularly those commonly classified as being individualist or collectivist.

Equally interesting, however, is the second meaning, which carries the clearly evaluative, positive connotations of professionalism, i.e., communicating to a standard of excellence which is recognized as exemplary within a field. Even though this chapter espouses a less elitist notion of professionalism, the roots of professional communication are clearly in elitism.

The source of this excellence lies in three aspects involved in the exercise of any profession:

[1] My use of *professional communication* here excludes much of what is commonly understood by the term, i.e., the communication practices of the traditional professions. Nonetheless many of the points raised in this chapter derive from analysis of these domains and may have implications in these settings.

1. mastery of a body of knowledge;
2. mastery of skills; and
3. the responsible use of this knowledge and these skills, i.e., understanding of, and commitment to, professional ethics.

On this basis, we can formulate a first proposition:

1. *The professional applies a body of knowledge by exercising a range of skills in an ethical manner.*

These three aspects—knowledge, skills, and ethics—will be discussed further below on the basis of a brief exploration of the notion of *communities* or *Discourses.*[2]

PROFESSIONAL DISCOURSES AND THE DISCOURSE OF PROFESSIONALISM

Discourse Boundaries and Comparative Notions of the Person

The social analytical paradigm sees professionals forming and working within communities. These communities are constituted by Discourses, and as such are commonly referred to as discourse communities [22]. Members of such communities have command of a discourse system, marked by specific socialization procedures, face systems, forms of discourse, and ideologies [4]. Professions are discourse communities into which novices are acculturated through a movement from exclusion to inclusion. The process is a long and painful one, requiring great personal sacrifice and material expense. Becoming a professional involves rites of passage. It involves acquiring the outward trappings of identity as well as the personal qualities characteristic of the target group. The professional communicator is likely to be a more or less central member of several such groups.

But Discourse boundaries are also associated with other groups, such as ethnic cultures, organizations (corporate cultures), age, and gender groups. We all communicate as members of a range of such communities. Uniquely, however, the communication professional has a specific responsibility for communicating *across* these boundaries, to members of other discourse communities, and in particular to non-professional communities. In other words, professional communicators are constantly face-to-face with the management of difference. This is emerging as one of their main functions in the post-industrial societies, to be

[2] In this chapter, following Gee [22, pp. 143-145], I use the terms *Discourse* (countable, with a capital D) and *culture* synonymously, and alternate freely between *interdiscourse* and *intercultural* communication. Focusing on discourse is analytically more exact, but culture is clearly the more familiar term.

dedicated managers of diversity, serving as "a facilitator, and as a catalyst for contact and change within and between cultures" [23, p. 295].

When communication occurs across significant Discourse boundaries, a local mode of discourse arises known variously as an *interculture* or a *third culture* [23]. An interculture is characterized by dynamic, everchanging frames and norms developing out of the Discourses in contact. It is a local system of relatively unpredictable communication since participants tend not to follow the behavior patterns characteristic of their own communities. Rather, they adapt their behavior toward what they conceive may be more acceptable to their interlocutors, but this very adaptation may make interpretation more problematic than the easily-read behaviors expected as typical of a certain group [4]. In situations of mutual cooperation, adaptation, and respect, the interculture can be a positive environment contributing the benefits of cultural synergy to individuals and organizations [24, 25]. The converse also holds. Cultural differences have great destructive power. The professional communicator is radically, therefore, an effective manager of diversity, wise in the ways of intercultures.

Learning to work within intercultures requires not merely a technological change, the adaptation of one's manner of speech or one's outward politeness to those of others. Differences in discourse practice reflect fundamental differences in world view and in the conceptualization of the person, a link that has been described by S. Scollon [26] drawing on the analyses of metaphor of Carbaugh [27] and Lakoff and Johnson [28], and the intercultural analysis of self by Hsu [29]. The current renewal of interest in the self, and deepening awareness of the differences in self-views between cultures, has led to a reevaluation of the notions of identity and the relations between identity, discourse, and literacy, supported by studies of their ontogenesis in child rearing practices using techniques of interpretive developmental psychology (for a survey of this work see [30]; for studies of self-construction through narratives see [31, 32]). Discourse practices simultaneously reveal and mould self-concepts, much in the same way as communication experiences constantly define and develop schemas, scripts, and expectation frames for interactions [33, 34]. Changes in communication practice thus have radical implications for speakers' self-concept, and surface features of discourse may, iceberg-like, rest on metaphorical depths of identity.[3]

Standard professional and business communication textbooks such as those mentioned above assume the notion of an individual, of the self, and an associated ideology of choice, freedom, and mutual respect for differing views. The

[3] As an example of how an apparently technical, surface feature of discourse can rest on deep conceptions of identity, see the analysis of plagiarism in the academic work of Hong Kong Chinese students in [35]. Here, difficulties in adopting Western academic referencing conventions is traced to contrasting views of responsibility for, and ownership of, discourse.

unmarked values are those of the Utilitarian Discourse System [4, 36], [illegible] of United States public discourse as described by Carbaugh, whose analysis I shall frequently cite in this chapter. Essentially this Discourse metaphorically constitutes the self as a container, an individual who "has" such a self, and assumes an ethical position that values independence, awareness, and self-display through communication. This ideology involves its own paradox, succinctly expressed by Carbaugh: "The cultural preference for choice itself requires conformity to a code of the individual where life is put into individual terms, thereby silencing the sociocultural forces in the making" [27, pp. 48-49]. The espousal of individual freedom, and the moral assertion of maximizing choice as the greatest good, involves its own contradiction in that it restricts the freedom of those for whom individualism is not the preferred world-view and for whom concepts related to the self are less salient, cultural categories than concepts related to social collectives.

This contradiction is at the heart of a dilemma for professional communications trainers preparing students for the international workplace. Given that any Discourse involves ideology (including a notion of the person), and professional communicators are essentially managers of ideological difference, what ideology are we to teach? Can we be content to teach, through standard, skills-oriented textbooks, an individualist ideology that superficially espouses diversity but through its focus on individual responsibility excludes those for whom the collective is a dominant component of the person? This is to assume what Carbaugh calls "a tyranny of openness" [27, pp. 111-112]. For communication teachers working with students in an East Asian setting, this question is unavoidable. But given the internationalization of business, the question must be considered by all professional communication teachers whatever their situation.

An answer to this dilemma is suggested by Cheney and Vibbert when they write: "To manage one's audience in discourse is also to manage one's identity in discourse, whether that 'one' be an individual or some kind of group" [37, p. 185]. Professional preparation must be founded on an understanding of comparative conceptions of the person, explored through the practice of interculture management skills and flexible identity management. It is especially through the analysis of ideology in discourse that both students and teachers are able to uncover the values rooted in the socially dominant Discourses they encounter first in the academy and then in the workplace. Teaching a multiplicity of Discourses is impractical. Teaching the analytical skills, the heuristics, and hermeneutics for recognizing the ideological stands of one's self and others, is a valuable educational endeavor.

We can sum up these complex issues in a deceptively simple proposition:

2. *The professional is able to communicate within and across Discourse boundaries.*

Discourses, Power, and Technology

In addition to these implications for personal identity, communication across Discourse boundaries, and particularly communications between professionals and the general public, has become prominent as a major social and political issue. Professionalism incorporates a status achieved through acquiring knowledge and skills, a status that wields organizational and social power through the command of Discourses. Critical analyses of professional Discourses, frequently appealing to Foucault for authority, reveal the workings of this power and call on professionals to become aware of these mechanisms in their communication particularly with those outside their communities [38, 39].

The power relations constructed and maintained by medical and legal discourse (the prototypical professional Discourses), which have been widely analyzed, cannot be seen as value-neutral. This analysis reflects a broad social change in power relations away from these professional groups toward a technological dominance, partially apparent also in the technologizing of the medical and legal roles. One of the irrational triumphs of rationalism has been the transformation of the dominance of church and state into a dominance of a Technological Discourse, one aspect of the Utilitarian Discourse System, currently reaching a peak in the movement toward and beyond total quality management. For professional communicators, command of this Discourse is a particular source of social power: the ideology of professionalism is intimately tied to command of both technology and its associated forms of discourse.

Technological discourse, however, has been accused of kidnapping rationalism in modern organizations:

> exclusive attention to economic and technological development serves as a form of domination and narrows the domain of reason in organizations. As an ideology, technological thinking imposes a singular standard for good reasons in organizations. Technological thinking is particularly insidious because of its presumed neutrality. One task of critical research is to expose its purposiveness and to recover reason from its sway. This is the task of a communication researcher [38, pp. 167-168].

In the West, Utilitarian Discourse is now a dominant system widely presumed to be value-neutral precisely because of its dominance. It is the status quo and the touchstone by which other value systems are defined. An additional characteristic which supports its position is its preference for a simple, direct, unadorned style, "a rhetoric of non-rhetoric . . . a discourse which persuades by appearing not to be trying to persuade" [36, p. 30]. Barton and Barton, in their analysis of maps as ideological representations, make a point that is applicable to techno-utilitarian discourse in general: it is "a discursive mode that . . . *naturalizes* and universalizes a set of practices so that the phenomenon represented appears to be described rather than constructed" [40, p. 53].

The dominance of Utilitarian Discourse is linked to the emergence of a technocracy, a new elite of the technologically adept. Business communication textbooks, along with espousing the ideology of the individual, contribute to these new power relations by presenting this Discourse as their explicit norm. As we have noted, however, professional communicators cannot work solely within a single Discourse; this would involve the unreflecting exercise of power through this Discourse and an inability to perform adequately in intercultures. Once again, educators are drawn to the need to base preparation programs on development of the analytical skills necessary for developing awareness of the roots of power in discourse.

Summing up the issues thrown up by this brief analysis, we can propose:

3. *The professional is able to reflect critically on the workings of power through discourse.*

PROFESSIONAL KNOWLEDGE

It is a cliché that knowledge is power, and the acquisition of knowledge has always been a key element in programs of professional preparation. The body of knowledge appropriate to the preparation of communication professionals is constantly being redefined. For curriculum development purposes, it is convenient to distinguish three categories of knowledge: technical, contextual, and general (or world) knowledge.

By *technical knowledge* I mean the concepts and information that professionals have access to as the knowledge base of their particular discipline. Doctors study anatomy and pharmacology: lawyers study tort and constitutional history. Professional communicators study the disciplines specific to the field of professional communication. These include organizational communication, organizational behavior, media theory, intercultural communication, cross-cultural (comparative) communication, conversation, discourse and genre analysis, ethnolinguistics, and pragmatics in monocultural and intercultural interaction.

Professionals also need *contextual knowledge,* knowledge of the contexts in which communication takes place. For communication professionals, this implies knowledge domains relating to understanding public and private organizations, business practices, and the sociopolitical context. These cannot be limited to the domestic environment but must extend, as far as is feasible, to understanding the business, social, and political backgrounds of interlocutors.

Given the professional communicators' contacts across Discourse boundaries, a broader, *general knowledge* is also essential. This domain includes historical, geographical, and more broadly-termed "cultural" knowledge. It cannot be easily taught, but is experientially acquired. In this case there is no substitute for voracious reading and intensive broadbased socializing, both of which must be encouraged by any communications program!

Mastery of knowledge raises further issues. Traditional models of teaching and assessment (lectures, tutorials, essay-writing, and examinations) assumed the acquisition of knowledge in the sense of an internalized, memory-retrievable set of integrated concepts, information, and techniques. The explosion of available information and retrieval tools has made this model obsolete. Mastery of knowledge has to be extensively supplemented, or even largely replaced, by *access* to knowledge, with the appropriate teaching of *access skills.* These knowledge acquisition systems (to borrow a concept from expert systems development) include classic study skills and research methods, Internet contact and search skills, and the broad range of quantitative and qualitative research techniques that organizations use to monitor their environment.

Proposition four summarizes these issues.

4. *The professional has access to the technical, contextual, and world knowledge necessary for communication within the communities concerned.*

PROFESSIONAL COMPETENCE

Knowledge is complemented by skills. Whereas acquisition of information requires the use of access skills, knowledge is manifested by performance based on the exercise of analytical skills. The professional communicator, on the basis of knowledge of the fields specified above, is specially equipped to analyze communicative interactions. This analytical ability is demonstrated by superior performance in intercultures, in other words by discourse practice. Discursive ability is therefore the observable outcome of the acquisition of knowledge and analytical abilities. (I will not enter the fascinating but futile debate on whether acquisition of discourse *is* the acquisition of knowledge!)

This discursive ability I class as a domain of performance skills employing the communicative resources of communities, i.e., codes and genres. These resources are culturally heterogeneous and constantly developing, and are employed by the practitioner to a high standard of excellence.

Codes and Genres

Codes include primarily languages, paralanguages, body languages, and visual rhetoric including document design and the codes involved in the fine and graphic arts. For the international professional communicator, language codes rank most prominently, and of these primarily English as a world language. Whichever language is the focus, both general and specific codes have to be considered: the professional requires command of whichever Languages for Specific Purposes (LSPs) apply in the discourse communities concerned.

Discourse skills, employing languages, and other codes, are highly specific, including, for example, sophisticated communication tasks such as media presentation, press reporting, cross-cultural negotiation, or corporate identity packaging for diverse audiences.

Codes as systems are abstractions from the specific genres in which they occur. Genres are not simply conventional arrangements of a particular code (e.g., a particular language). A focus on language use, based on a limited, linguistic interpretation of what a "text" is, gives only an impoverished view of genre. Rather, genres are formed at the intersection of conventional concurrent uses of a variety of codes. For example, the business letter as a genre incorporates conventional uses of codes such as language, document layout, typography, physical aspects (e.g., paper quality), and time related codes (promptness of reply, conventional times of issue, etc.). Genre analysis reveals the conventional purposes, structures, forms, and uses of these multiple codes as communicative events within communities.

When the strategic use of genres is considered, we enter the realm of rhetoric. Conventional genres and rhetorical strategies can be both liberating and constraining; they provide the possibility for communication to take place, while at the same time limiting the range of this communication. Genres in the communicative arts are as complex and subtle as genres in the fine arts, with creativity based on generic mixing, embedding, and subtle intergeneric allusions [41-43].

Professional texts are also becoming increasingly complex, and their production a matter less under individual control and increasingly team- and organization-based. This is particularly the case in much technical writing [13, 44, 45]. This development further validates the analysis of professional communication as social enactment rather than individual expression [46]. The professional communicator and the communication manager therefore require the ability to manage the production of complex genres, and training should include publication project management skills. These include management of spoken events (series of meetings, briefings, presentations, etc.), use of computer-based document management systems, and contributions to the design of information systems in general. Publication projects can be conceptualized as a sequence of communicative events (interactions with writers, graphic artists, proof-readers, etc.) involving written and spoken texts, leading up to the production of a final, target text, or set of texts, which are at the heart of their own events (e.g., consumers exposed to a marketing campaign, technicians referring to a set of manuals).

International communication training is complicated by the fact that genres are Discourse-specific; they are non-isomorphic, varying in form and function between discourse communities. A common generic name in English often conceals such underlying differences. A typical departmental meeting in a Chinese corporation may well have an annunciatory function; its counterpart in the United States, at least according to the textbooks, would involve argumentation, decision-making, consensus-building, and explicit statements of personal commitment. It is clear that training programs must raise awareness of generic differences in order to prepare students for participation in groups operating within an interculture.

To make things more complex, codes and genres are never static. They develop, and to a limited extent, can be developed especially by professional associations. Professionalism, therefore, also involves commitment to a proactive development of existing practice. The professional, both in training and practice, cannot be content to maintain discourse as is, but is always engaged with what will be and what should be.

These dynamics again direct professional preparation toward the teaching of heuristics and hermeneutics rather than performance, toward strategies for analyzing communication problems rather than solutions, toward the teaching of genre analysis rather than genre reproduction, and communication task analysis rather than textual modeling. Unfortunately, most undergraduate communication texts still teach "a right way" to run meetings, a "standard form" for the business letter, an "appropriate strategy" for the bad news message. All these approaches assume a homogeneous discourse community and do a disservice to professionals for whom *no way* will be *the right way* in all the disparate situations they will meet.

Standards of Performance

Finally, codes and genres are used with varying degrees of competence. Performance standards correspond to the norms of interpretation characteristic of certain discourse communities [47]. Excellence in communication is a question of community norms. It seems, then, that if we want to know about acceptable standards in business letter writing we go to the relevant business community; if we want to know about acceptable standards in copywriting we go to the advertising industry and to consumers. In short, the arbiters of quality in professional communication are the stakeholders in that communication.

Unfortunately, reflection from an international perspective muddies this simple view. It fails to comprehend that the locus of responsibility for creating understanding varies between cultures. In the Western rhetorical tradition, rooted in democratic political persuasion, responsibility for adjusting the message to meet the needs of the audience falls firmly in the camp of the speaker. Thus the politician is judged by the people, the teacher is judged by the student, and the technical writer conducts audience analysis before conceptualizing a rhetorical strategy. This assumption underlies most undergraduate communication texts.

In East Asia, the responsibility is not so clearly positioned. Not only is the prototypical public discourse an expression of a consensual position rather than an individual persuasion, the speaker is also granted more autonomy and the listeners placed in a more dependent position with greater responsibility for achieving their own understanding of a more or less recalcitrant text [48]. The right to speak is inseparable from the hierarchical position of the person; in other words, any decision to speak involves assuming a hierarchical position, and typically the firm pronouncing of an opinion marks either superior status or representation of a group position. To draw a familiar example from education, student-centered classroom practices imported from the United States and elsewhere pose rhetorical problems in cultures which place the locus of communicative control on the teacher. Attempts to impose interactive learning practices on Chinese students may be just another example of Carbaugh's "tyranny of openness," an attempt at exporting an alien ideology that has been criticized in another form by Third World women writers on U.S. feminism (e.g., at the United Nations Fourth World Conference on Women held in Beijing in September 1995). In corporate communications too, the appropriateness of involving, or appealing to, stakeholders (such as consumers) for assessment of the quality of communications cannot be assumed.

In short, the quality of professional practice cannot be judged simply by stakeholders. Different cultures impose different "rights to speak" on different groups and the most powerful stakeholders, the opinion leaders, do not occupy equivalent positions in all communities. Professionals under training must therefore study *comparative* rhetoric, not only to develop appropriate rhetorical strategies for use in intercultures but also for an understanding of comparative notions of quality.

Proposition five summarizes these issues.

5. *The professional creatively uses the communicative resources of discourse communities to a standard that is recognized as exemplary.*

PROFESSIONAL ETHICS

Ethical Debate as Discourses in Contention

In the West today, ethics is news in a way it has seldom been since the theological debates of the seventeenth century. The ethics of public and private life came alive as an issue in the United States in the 1960s and has not died since. This ethical consciousness has taken many forms including anti-discrimination movements, the women's movement, the investigations into transnationals (ITT in Chile, and the bribery scandals leading to the Foreign Corrupt Practices Act of 1977), the post-Watergate investigations of illegal campaign contributions [49, p. ix], and the public fascination with the sexual practices of the great and supposedly good. Clearly, professionals apply their knowledge and skills in ways that involve the real use of power to the benefit of identifiable groups and individuals. Inevitably, their communicative acts have an ethical dimension. Inevitably, too, comes ethical conflict. Consideration of ethical issues in both business and the media, particularly in relation to news, advertising or entertainment, must be included in the training of the professional communicator.

The media spotlight illuminates personal and corporate rhetoric as ethical action. The current debate over political correctness is essentially about communication ethics. In this debate, traditional ethical values such as truth, honesty, and decency provide one voice. As a counterpoint, we hear a social constructionism that views society and societal roles as oppressive, restrictive of personal freedoms, and limiting the personal right to a free building of an identity through the exercise of choice. This new ethics values independence, awareness of issues, and communicative openness [27, pp. 62-69]. Against this has arisen a libertarian view that sees society as a repository of ethical codes serving as a protector of the individual and of women in particular [50]. Though usually couched in universalist terms, these positions are not easily exported. In particular, they make uneasy companions with East Asian norms, which have tended traditionally to value interpersonal dependency and restraint in communicating personal opinions.

These ethical choices and disputes reflect larger social movements of Discourses in contention. A decision toward action in any one case represents small victories or defeats for these discourse systems. The Utilitarian Discourse referred to above, which places effectiveness as its supreme value, though it may be dominant in the business world, is by no means secure in its hegemony. It is constantly challenged by the competing Discourses valuing individual or group rights, global responsibilities (such as the ecological consciousness), or the moral universalism of organized religious groups. This contention demands increasingly explicit expressions of ethical positions by individuals, professional associations, and corporations.

Codes of Practice and the Personalized Corporation

In the traditional professions, ethical issues were codified in professional codes of practice. As Behrman points out, these codes formalized standards for the provision of services quite unlike other services rendered by business. The professional associations instituted peculiar practices such as differential pricing of identical services according to the client's ability to pay, consistent standards of practice whatever the situation, and setting of prices with reference not to quality but to time on task. Such practices formed the core of professional ethics, but ironically many are forbidden to the new professionals. Discriminatory pricing, for example, is specifically prohibited in U.S. business law [15, p. 101]. These practices are therefore not carried over into the new professions, nor necessarily to the new forms of the old professions.

Rather, there has been a carry-over of the traditional *forms of discourse* in which ethical constraints on professional action have been voiced. There has been a proliferation of codes of conduct, which are now issued by professional associations, corporations, occupational, and political groups as part of the professional coming of age referred to previously. An occupational group that adopts a code of practice marks itself off as a profession, and subscription to the code is one method for defining the boundary of the putative professional community. Proved infractions can lead to expulsion into the unprofessional outer darkness, but enforcement requires surveillance and sanctions, both of which are often lacking.

Corporations that adopt a code of business practice claim allegiance to public Discourses espousing social responsibility. But these corporate (as opposed to professional) codes of practice pose a special ethical problem relating to the professional identity that is the focus of this chapter. Their development marks an interesting movement away from the individual ethical and legal responsibility typical of professional codes, toward a conception of corporate responsibility that can act as a shield for individual malpractice. The ethical situation is confused by the peculiar status of corporations, which are established, in the United States, as "a person in the eyes of the law" [51, p. 97], one example being *First National Bank of Boston v. Bellotti* by which the Supreme Court in 1978 established free-speech rights for corporations.

This legal re-definition of groups as individuals is matched by a clear tendency in both ordinary speech and academic analysis to speak *metaphorically* of the corporation as a person. Velasquez attacks such personalization of the corporate group, spread for example through the metaphor of corporate "health" and organic models of corporate development, as a new form of totalitarianism [52]. He argues that the only sense to be made out of the notion of a corporation being morally responsible and subject to blame and punishment lies in an ultimate reduction to individual responsibility. The trend toward the personalized corporation confuses the ethical responsibilities of the individual as opposed to those of the organization [53, pp. 205-209]. Focusing on corporate responsibility is dangerous, especially when combined with the view (associated with Milton Friedman [54]) that the only responsibility of the corporation is to make profits.

Codes of practice do, however, embody the recognition that ethics cannot be seen solely from the narrow perspective of any one community, nor from the point of view of individual rather than collective action. The practices of professional groups, like those of corporations, are very much the concern of other, wider groups. Professional communities, and increasingly corporations, find it socio-politically and to some extent economically advisable to present themselves as responsible to their stakeholders, including the wider communities in which they operate. Ethical codes of specific groups therefore serve to protect members of these communities, though they may also be manipulated to protect members of the professional community involved. This "stakeholder approach" in the analysis of business ethics is increasingly popular [55].

For international communicators, this implies a need to take up, and somehow to reconcile, local, national, and even global perspectives on ethical questions. Simple attributions of cultural values (such as collectivist vs. individualist) are not much help to us here. For example, the United States notion of the corporate person reveals a supposedly individualist society in a strong assertion of collective responsibility. In contrast, consider a recent statement attributed to Chinese prosecutors in an enquiry into a factory fire in Shenzhen, PRC:

> We should not only judge by terms in the contract. We can be sure that it is actually a factory owned by [the accused] and he should be held responsible for every aspect of factory management [56].

Here a society stereotyped as collectivist adopts a strongly individualist position on personal responsibility. We must look beyond broad cultural typologies for insights into reconciliation of ethical conflict.

Ethical Reconciliation as a Meta-Discourse

These ethical issues raised by corporate and personal rhetoric again raise for the professional communicator the problem of the multiplicity of communities. Corporate ethics require the reconciliation of diverse ethical viewpoints held by the multiple groups who are stakeholders in the corporate enterprise, including managers, stockholders, competitors, labor organizations, governments, consumers, and communities domestic and overseas [15, p. 242]. Porter usefully interprets ethical corporate writing as involving "a sense of the constraints and guidelines" derived from tradition and culture, disciplines and professions, communities, and corporate policies and practices [57, p. 142]. Unfortunately, he gives no hints as to how this reconciliation can be achieved. Similarly, on a personal level, ethical communication requires reconciliation of competing views, including those of libertarians and the espousers of political correctness.

The implication of a discourse-based analysis of ethical issues is that training programs should prepare students in two ways. First, students must be trained to recognize the ethical values embodied explicitly and implicitly in competing Discourses, for example, by analysis of corporate codes of practice, public statements in crisis management cases, consumer reports, environmental lobbyist, or stockholders' reports. Wherever possible these should be drawn from diverse cultures, countries, and industries. These should be treated as case studies for discourse analysis rather than subjects of moral reasoning for individual choice, which is a pedagogy further entrenching the Western notion of the professional as an individual self. This understanding should be complemented by a productive command of rhetorical strategies in the composition of texts embodying a variety of ethical stances.

Second, any program should attempt to develop, for its own context, a meta-Discourse of ethical reconciliation. As an interdiscourse, the characteristics of this meta-Discourse are hard to predict. The meta-Discourse will share features from the students' and lecturers' home cultures, the academic culture of the institution and the analytical frameworks used in the course. Fostering this constructive interdiscourse is an important aim of program management in much the same way as communications management fosters a communication climate within the corporation.

Proposition six summarizes these ethical issues.

6. *The professional applies knowledge and skills in accordance with codes of conduct that are recognized as ethical by communities.*

PROFESSIONALS IN ORGANIZATIONS

Despite its Western roots, the stereotyped professional is not conceptualized as an independent citizen. The psychology of professionalism is not the psychology of the isolate individual. Neither is it, on the other hand, the psychology of the individual in society. Rather it is the individual in the organization. The stereotype presents the professional as acquiring identity through identifying, by individual choice, personal needs with the needs of organizations. Only outside the organization does the professional transform back into an individual, stripped of the trappings of position, the logos and corporate identities, the departmental agendas and group alliances. There is no such thing as the professional at home; rather, the professional transforms home into an office. The professional inhabits not the natural world, but the human-made world. Despite occasional team-building excursions into the wilderness, the true haunt of the professional is the corridors of power. To understand the nature of the professional we must look at the ecology of the organization.

Professional Purposes and Effectiveness

Organizations are essentially goal-driven, being established formally at a particular time and place in order to achieve certain ends. These goals are very often articulated in articles of incorporation, or in corporate mission statements. Within the organization, sub-groups with their own agendas, contributing more or less directly to the overall organizational mission, demand competing loyalties from their members. Professional activity must therefore be seen as effective at multiple levels. It is purposeful in achieving the aims of the organization as a whole, as well as the aims of sub-groups and individuals within the organization.

This word *effective* is central within the professional stereotype. It calls to mind the ruthless effectiveness both endorsed and satirized in films such as *Wall Street* and *Working Girl.* This clearly calls for a caveat. In training for the international arena, we have to avoid a glib fallback on what seem to be increasingly outdated, ethnocentric ideas of *effective management.* Purposes of direct relevance to the organization as a whole must be maintained in equilibrium with purposes relating to work groups, to family and social groups, and to society at large. These domains of purpose are not necessarily consistent with maximizing organizational effectiveness in terms of market success (see the comments on the Utilitarian Discourse above). Different working environments demand different balances of priorities between these domains. In some cultures, the maintenance of harmonious relations within a work group may be accorded a higher priority than task achievement. The maintenance of workplace harmony is a key element in Chinese management. Scandinavian high scores on Hofstede's femininity dimension indicate that a key occupational motivator is to maximize the quality of family life [25, 58]. In other words, the bottom line is not universally perceived as the fundamental measure of organizational effectiveness. Though necessary, it may not be sufficient.

This comparative understanding of effectiveness is implicit in Hofstede's more recent work on the intercultural limits of management theory [59]. We can conclude that the professional communicator must be able to balance the demands made by cultural traditions on the individual in the workplace with the sometimes conflicting demands of personal, familial, occupational, and social development. Bearing in mind this comparative notion of effectiveness, proposition seven states:

7. *The professional is effective in reconciling and achieving multi-level purposes through communication.*

Management of Organizational Communication

Earlier in this chapter I outlined the emerging role of communication managers. This role relates to both internal and external corporate communication. Internally, it involves developing corporate communication practices directed toward fostering a positive communication climate and organizational culture (see the discourse approach deriving from Clifford Geertz [60] surveyed by Smircich and Calás [61]). Externally, it relates to developing corporate image and identity through marketing, public relations, and issue management.

Communication climate, analyzed at organizational, group or dyadic levels, is a complex construct with established effects on group development and productivity. It is conceived as a cognitive map or schema employed in interpreting messages received in an organizational context. Its main determinants are leadership styles and network position. Falcione et al. provide an overview of studies in United States organizations which strongly suggest that:

- member satisfaction increases to the extent that leadership style allows for participation and equal distribution of control across participants . . .
- perceived position within a group of message senders/receivers significantly affects affective response to the group and performance with it [62, p. 209].

They identify further factors contributing to positive communication climate, including communication accuracy, openness, warmth, and task facilitation.

Findings such as these, which identify specific contributory values, are not necessarily generalizable to other cultures. In contrast, systemic findings concerning the determinants of communication climate may well be generalizable. Leadership style and network position may well be universal determinants, though specific values of these variables will have different effects in different environments.

Further findings indicate that it is the perceived (rather than the objective) presence of a factor such as network position that constitutes the controlling variable [62, p. 215], though it is not clear whether differences in perception of organizational communication are due to individual, group, or environmental differences. Communication climate is also created by a combination of both ambient (unconscious, subliminal) and discretionary (consciously perceived, manipulated) stimuli ([63], cited in [62, p. 218]). What does seem clear, then, is that the communication manager must be concerned with both actual communication structures and structures of the imagination. Managing corporate communication climate is like administering justice: appropriate communication procedures must not only be done, they must be seen to be done.

Training programs need to present the tools available to the communication manager, even though these appear crude compared to the complexity of the material to be worked. Managerial manipulation of organizational climate, so-called "cultural engineering," has limited and unpredictable effects. Investigative approaches include communication audits and communication needs analyses. Instruments of development include tasks traditionally falling under human resource management, such as the design of employee handbooks, policy and procedures manuals, corporate style manuals, house organs, staff newsletters, etc. Information systems management has further implications for communication climate, as do the management of change generally, the introduction of technological innovations, restructuring, and the associated training processes.

The external responsibilities of communication managers include the public communication of organizational image, identity, and values in relation to significant issues (see [37] for definition of these terms and a survey of the literature). The tools here are those for monitoring the environment, both physical and symbolic, such as testing for brand recall, advertisement recall, or public relations auditing. Tools for enactment are the huge variety of persuasive campaign methods attempting to mould public opinion, to " 'locate' the corporation in the domain of public discourse" [37, p. 184].

In an international program, these issues and tools must be presented from a culturally comparative perspective. There is an established literature on the management of international work-groups and intercultural negotiation, a growing body of research on international marketing, but very little work as yet that takes a discourse approach focusing on comparative rhetoric in the use of English as an international language of business. Small-scale research into local uses form excellent project work for professional communication students while advancing the field.

Finally, in this section, we should note that the internal-external division employed here is somewhat arbitrary. Professional communicators not only communicate across corporate and departmental boundaries, their communication also defines where those boundaries should be set. Clients may contribute more to organizational development than many employees. For large corporations, employees may be the best publicists, their informal private contacts with the general public having more effect than PR campaigns. Trainees must also, therefore, be able to combine systems thinking that can treat organizational boundaries as mutable with a discourse approach that conceives of those boundaries as constantly constructed, and at times contested, in a variety of corporate encounters.

To summarize:

8. *The professional is able to manage internal and external organizational communication by defining and implementing communication policy.*

PROFESSIONAL DEVELOPMENT

We live in a working world subject to accelerated development. In many regions (including East Asia) this acceleration is acute, extreme, and involves dislocations of traditional societies and development of new economic, political, sociocultural, and technological orders. The professional communicator, though a creator of new orders, is also very much at the mercy of uncontrollable forces of change. The agents of change must constantly change themselves or risk becoming marginalized.

Continuing professional development is essential in response to constantly changing global and local conditions. Was it ever true that a professional would receive training as a practitioner and then be qualified for a lifetime? Currently, constant renewal is the norm and there is a terror of stasis (stagnation as opposed to stability), a position not unconnected with the imperative noted by Carbaugh toward constant personal growth and development through individual choice-making [27, p. 56]. Again we are forced to the conclusion that professional training, as any educational endeavor, must focus on heuristics, on skills for constantly developing knowledge and skills, rather than on mastery of a fixed body of knowledge or range of ability.

As a final proposition, therefore, I would like to suggest:

9. *The professional continuously seeks to develop through active reflection on performance.*

PROFESSIONAL COMMUNICATION: TOWARD RECONCILIATION AND SYNTHESIS

Any learning is a becoming. As we learn, we change ourselves. Learning is the development of new identities, and the more this learning involves communication the more intimate is its involvement with the person. If learning to communicate is the construction of identity, of identities, then it is imperative that the communications teacher understands the goal of this becoming.

This chapter attempts to address the question of what professionalism can mean in the context of international professional communication. It attempts to lay a foundation in discourse studies on which such training can be based, an approach that will help novices and established professionals to think through communication issues within multiple discourse frameworks in order to reach the best compromises they can. By analysis of the current stereotype, professionals can be brought to understand the limits of professionalism, its dangers as well as its potentials.

This approach also provides a direction for educators designing programs of professional preparation. Against the image of the professional as an expert user of the Utilitarian Discourse System, it sets a redefinition of a professional as an expert reconciler of difference, as manager of intercultures, whose fundamental function is the creation of synthesis out of multiple Discourses. We are exploring the practical educational consequences of this position in the BA in English for Professional Communication running in the Department of English at the City University of Hong Kong.

This chapter further suggests that professional development is not a question of the acquisition of knowledge and skills, not even of discourse skills, but of the building of a new, fluid identity. It suggests that professional communication consists of the management of identity through communication within intercultures whose characteristics can only be partially predicted. The logic of this approach leads us toward the teaching of heuristics and analytical skills. It also raises questions about personal and group identity often at a period in students' lives, and in the life of their countries, when such issues have an intense salience. The approach thus poses unique challenges for learners.

The approach also requires communication trainers to remember that the teaching enterprise is itself an example of professional intercultural communication and that the classroom is itself an interculture. Teachers are professional communicators too. We must therefore constantly analyze the discourse of the classroom, and ensure our students are also able to analyze it. Our training must be a microcosm of the business world. Not in the sense that the academy should be run at a profit. Not even in the sense that we should teach the practices and values of the business world. Rather, the strategies for successful intercultural professional communication in the business world should be constantly demonstrated and modeled in the classroom, and the exercise of these analytical abilities should be an important criterion for success in academic tasks. This perspective provides daily personal and professional challenges for teachers equal to those faced by the students.

The Greek robber Procrustes had two beds, one long and one short. He would stretch his short visitors to fit the long bed, and cut off the limbs of his tall victims to fit the short one. International professional communication is a Procrustean bed that few can sleep on without some arduous extension or excision of their identities. We have no choice but to occupy it as best we may. No Theseus, no champion of simplicity, can return us to the monocultural certainties of earlier generations.

ACKNOWLEDGMENTS

I am particularly indebted to Drs. Ron and Suzie Scollon whose published writings and corridor conversations are the source of many of the ideas in this chapter. I would also like to thank Steve Walters and Pam Rogerson-Revell whose co-teaching on the course *Fundamentals of Professional Communication* prompted much of the development of this chapter. Also, of course, my thanks are due to the students in the BA(Hons) in English for Professional Communication, City University of Hong Kong, whose confidence in the concept demands a considered account of professionalism, toward which this chapter is hopefully a contribution.

REFERENCES

1. L. Putnam and M. Pacanowsky (eds.), *Communication and Organizations: An Interpretive Approach,* Sage, Newbury Park, California, 1983.
2. L. Odell and D. Goswami (eds.), *Writing in Non-Academic Settings,* The Guilford Press, New York, 1985.
3. N. Blyler and C. Thralls (eds.), *Professional Communication: The Social Perspectives,* Sage, Newbury Park, California, 1993.
4. R. Scollon and S. Wong Scollon, *Intercultural Communication: A Discourse Approach,* Blackwell, Oxford, 1995.
5. C. Thralls and N. Blyler, The Social Perspective and Professional Communication: Diversity and Directions in Research, in *Professional Communication: The Social Perspective,* N. Blyler and C. Thralls (eds.), Sage, Newbury Park, California, pp. 3-34, 1993.
6. L. Barker and D. Barker, *Communication* (6th Edition), Prentice-Hall, Englewood Cliffs, New Jersey, 1993.
7. J. Forman and K. Kelly, *The Random House Guide to Business Writing,* McGraw-Hill, New York, 1991.
8. H. Murphy and H. Hildebrandt, *Effective Business Communications* (6th Edition), McGraw-Hill, New York, 1991.

9. R. Berko, A. Wolvin, and D. Wolvin, *Communicating: A Social and Career Focus* (5th Edition), Houghton Mifflin, Boston, 1992.
10. C. Bovée and J. Thill, *Business Communication Today* (3rd Edition), McGraw-Hill, New York, 1992.
11. H. Johnsson, *Professional Communications for a Change,* Prentice-Hall, Englewood Cliffs, New Jersey, 1990.
12. M. Hanna and G. Wilson, *Communicating in Business and Professional Settings* (3rd Edition), McGraw-Hill, New York, 1991.
13. T. Huckin and L. Olsen, *Technical Writing and Professional Communication for Nonnative Speakers of English,* McGraw-Hill International Editions, New York, 1991.
14. J. Swales, Discourse Analysis in Professional Contexts, *Annual Review of Applied Linguistics (1990), 11,* pp. 103-114, 1991.
15. J. Behrman, *Essays on Ethics in Business and the Professions,* Prentice-Hall, Englewood Cliffs, New Jersey, 1988.
16. T. Harris and J. Bryant, The Corporate Communication Manager, *The Journal of Business Communication, 23*:3, pp. 19-29, 1986.
17. T. Boswood, Employment Prospects for EPC Graduates in Hong Kong, in *Perspectives on English for Professional Communication,* T. Boswood, R. Hoffman, and P. Tung (eds.), Department of English, City University of Hong Kong, pp. 315-340, 1993.
18. B. Ticehurst, The Communication Manager: Organizational Role, Industry Needs, and Training, *Australian Journal of Communication, 17*:1, pp. 113-123, 1990.
19. S. Suchy, *Australian Communicators '88,* Centre for Applied Communication, Kuring-gai College of Advanced Education, Lindfield, 1989.
20. B. Ticehurst, G. Walker, and R. Johnston, Issues in Communication Management in Australian Organizations, *Australian Journal of Communication, 18*:3, pp. 81-97, 1991.
21. E. Goffman, *The Presentation of Self in Everyday Life,* Penguin, Harmondsworth, 1959.
22. J. Gee, *Social Linguistics and Literacies: Ideology in Discourses,* Falmer Press, Philadelphia, 1990.
23. F. Casmir and N. Asuncion-Lande, Intercultural Communication Revisited: Conceptualization, Paradigm Building, and Methodological Approaches, in *Communication Yearbook 12,* J. Anderson (ed.), Sage, Newbury Park, California, pp. 278-309, 1989.
24. N. Adler, Cultural Synergy: The Management of Cross-Cultural Organizations, in *Trends and Issues in OD: Current Theory and Practice,* W. Burke and L. Goodstein (eds.), University Associates, San Diego, California, pp. 163-184, 1980.
25. N. Adler, *International Dimensions of Organizational Behavior* (2nd Edition), PWS-Kent, Boston, Massachusetts, 1991.
26. S. Scollon, Metaphors of Self and Communication: English and Cantonese, in *Perspectives: Working Papers of the Department of English,* City University of Hong Kong, *5*:2, pp. 41-62, 1993.
27. D. Carbaugh, *Talking American: Cultural Discourses on DONAHUE,* Ablex, Norwood, New Jersey, 1989.

28. G. Lakoff and M. Johnson, *Metaphors We Live By,* The University of Chicago Press, Chicago, 1980.
29. F. Hsu, The Self in Cross-Cultural Perspective, in *Culture and Self: Asian and Western Perspectives,* A. Marsella, G. DeVos, and F. Hsu (eds.), Tavistock/Methuen, London, pp. 24-55, 1985.
30. S. Gaskins, P. Miller, and W. Corsaro, Theoretical and Methodological Perspectives in the Interpretive Study of Children, in *Interpretive Approaches to Children's Socialization (New Directions for Child Development 58),* W. Corsaro and P. Miller (eds.), Jossey-Bass, San Francisco, pp. 5-23, 1992.
31. P. Miller, R. Potts, H. Fung, L. Moogstra, and J. Mintz, Narrative Practices and the Social Construction of Self in Childhood, *American Ethnologist, 17*:2, pp. 292-311, 1990.
32. P. Miller, J. Mintz, L. Hoogstra, H. Fung, and R. Potts, The Narrated Self: Young Children's Construction of Self in Relation to Others in Conversational Stories of Personal Experience, *Merrill-Palmer Quarterly, 38*:1, pp. 45-67, 1992.
33. T. Van Dijk, *Text and Context: Explorations in the Semantics and Pragmatics of Discourse,* Longman, London, 1977.
34. T. Van Dijk and W. Kintsch, *Strategies of Discourse Comprehension,* Academic Press, Orlando, Florida, 1983.
35. R. Scollon, What is Plagiarism? A Note on Intercultural Problems of Identity in Discourse, in *Perspectives: Working Papers of the Department of English,* City University of Hong Kong, *5*:2, pp. 1-24, 1993.
36. R. Scollon, From Pidgin English to Professional Communication: English Teaching and the Utilitarian Discourse System, in *Explorations in English for Professional Communication,* P. Bruthiaux, T. Boswood, and B. Du-Babcock (eds.), City University of Hong Kong, pp. 21-40, 1995.
37. G. Cheney and S. Vibbert, Corporate Discourse: Public Relations and Issue Management, in *Handbook of Organizational Communication,* F. Jablin, L. Putnam, K. Roberts, and L. Porter (eds.), Sage, Newbury Park, California, pp. 165-194, 1987.
38. S. Deetz and A. Kersten, Critical Models of Interpretive Research, in *Communication and Organizations: An Interpretive Approach,* L. Putnam and M. Pacanowsky (eds.), Sage, Newbury Park, California, pp. 147-172, 1983.
39. N. Fairclough, *Language and Power,* Longman, London, 1985.
40. B. Barton and M. Barton, Ideology and the Map: Toward a Postmodern Visual Design Practice, in *Professional Communication: The Social Perspective,* N. Blyler and C. Thralls (eds.), Sage, Newbury Park, California, pp. 49-78, 1993.
41. E. Ventola, *The Structure of Social Interaction—A Systemic Approach to the Semiotics of Service Encounters,* Frances Pinter, London, 1987.
42. V. K. Bhatia, *Analysing Genre: Language Use in Professional Settings,* Longman, London, 1993.
43. V. K. Bhatia, Genre-Mixing in Professional Communication: The Case of 'Private Intentions' v. 'Socially Recognized Purposes,' in *Explorations in English for Professional Communication,* P. Bruthiaux, T. Boswood, and B. Du-Babcock (eds.), City University of Hong Kong, pp. 1-20, 1995.
44. Mary M. Lay and William M. Karis (eds.), *Collaborative Writing in Industry: Investigations in Theory and Practice,* Baywood, Amityville, New York, 1991.

45. R. Grice, *Technical Communication in the Computer Industry: An Information-Development Process to Track, Measure and Ensure Quality,* Ph.D. thesis, Rensselaer Polytechnic Institute, Troy, New York, 1987.
46. J. Comprone, Generic Constraints and Expressive Motives: Rhetorical Perspectives on Textual Dialogues, in *Professional Communication: The Social Perspective,* N. Blyler and C. Thralls (eds.), Sage, Newbury Park, California, pp. 92-108, 1993.
47. M. Saville-Troike, *The Ethnography of Communication* (2nd Edition), Blackwell, Oxford, 1989.
48. R. Oliver, *Communication and Culture in Ancient India and China,* Syracuse University Press, Syracuse, New York, 1971.
49. T. Tuleja, *Beyond the Bottom Line: How Business Leaders are Turning Principles into Profits,* Facts on File Publications, New York, 1985.
50. C. Paglia, *Sex, Art and American Culture,* Random House, New York, 1992.
51. J. DesJardins and J. McCall, *Contemporary Issues in Business Ethics* (2nd Edition), Wadsworth, Belmont, California, 1990.
52. M. Velasquez, Why Corporations are Not Morally Responsible for Anything They Do, in *Contemporary Issues in Business Ethics* (2nd Edition), J. DesJardins and J. McCall (eds.), Wadsworth, Belmont, California, pp. 114-124, 1990.
53. W. Shaw and V. Barry, *Moral Issues in Business* (5th Edition), Wadsworth, Belmont, California, 1992.
54. M. Friedman, The Social Responsibility of Business is to Increase its Profits, in *Essentials of Business Ethics,* P. Madsen and J. Shafritz (eds.), Penguin, New York, pp. 273-282, 1990.
55. J. Weiss, *Business Ethics: A Managerial, Stakeholder Approach,* Wadsworth, Belmont, California, 1994.
56. Ng Kang-Chung, Fire Trial Conditions Improve, *South China Morning Post,* p. 3, August 13, 1994.
57. J. Porter, The Role of Law, Policy and Ethics in Corporate Composing: Towards a Practical Ethics for Professional Writing, in *Professional Communication: The Social Perspective,* N. Blyler and C. Thralls (eds.), Sage, Newbury Park, California, pp. 128-143, 1993.
58. G. Hofstede, *Culture's Consequences: International Differences in Work-Related Values,* Sage, Beverly Hills, California, 1980.
59. G. Hofstede, Cultural Constraints on Management Theories, *Academy of Management Executive,* 7:1, pp. 81-93, 1993.
60. C. Geertz, *The Interpretation of Cultures,* Basic Books, New York, 1973.
61. L. Smircich and M. Calás, Organizational Culture: A Critical Assessment, in *Handbook of Organizational Communication,* F. Jablin, L. Putnam, K. Roberts, and L. Porter (eds.), Sage, Newbury Park, California, pp. 228-263, 1987.
62. R. Falcione, L. Sussman, and R. Herden, Communication Climate in Organizations, in *Handbook of Organizational Communication,* F. Jablin, L. Putnam, K. Roberts, and L. Porter (eds.), Sage, Newbury Park, California, pp. 195-227, 1987.
63. J. Hackman, Group Influences on Individuals, in *Handbook of Industrial and Organizational Psychology,* M. Dunnette (ed.), Rand McNally, Chicago, pp. 1455-1525, 1976.

PART TWO:
Rhetorical Strategies for the Global Workplace

CHAPTER 6

Intercultural Business Communication: An Interactive Approach

JÜRGEN BOLTEN

"COMPARATIVE" VERSUS "INTERCULTURAL" MANAGEMENT RESEARCH

Whenever we compare two phenomena, we can obtain successful results only if we consider them as independent items in their own context. The same is true when comparing cultures to find out how, for example, management styles, marketing strategies, or overall value systems differ from one another. What a comparison cannot do is to make a statement about what happens when, for instance, members of two different cultures interact. Generally, when individuals interact with members of another culture, they will not behave in the same way they would within their own cultures. They will adapt to some unknown degree, devoting more time to the other individual and thereby changing their expectations of what is considered normal. Individuals involved in this interaction create a new, "third" context, which is distinguished by interaction and process. The interaction, not the comparison, can be described as being "intercultural." This encapsulates the difference between comparative and intercultural management research.

Culturally oriented "comparative management research"[1] analyzes culturally independent aspects of business in selected countries. Many empirical studies

[1] See research reports in J. Henze [1, pp. 170-185], as well as N. Kumar [2, pp. 389-394].

from the past, such as those by Hofstede,[2] seem quite impressive even by today's standards, especially considering the amount of data examined. However, the tendency is to turn these results, representing a static comparison of two cultures, into formal guides for "correct" behavior vis-à-vis foreign companies or business partners. Clearly such findings, which refer to specific cultural groups at specific points in time, cannot automatically be applied to the dynamic situation of an interactive process between members of differing cultural groups. The value of knowledge based on contrastive studies is in no way lessened by this. However, it should be pointed out that this knowledge refers to a contrastive, *intra*cultural interaction [5, p. 75]. As such, it describes only a portion, albeit an indispensable portion, of the *preconditions* for the *inter*cultural process. What is missing is the interaction process itself that decides the success or failure of a strategy aimed at internationalization. Whenever this is ignored and the *knowledge* about a foreign culture is confused with the *action* in an intercultural context, then precisely what one would like to avoid occurs, namely, an increase in the formation of stereotypes and prejudice. Ever since the concept "intercultural" came into vogue in the late 1980s, especially through the humanities, many handbooks have appeared in the style of "Do's and Taboos in [. . .]"; similarly, management training programs[3] often make use of the label "intercultural," despite being based on the intracultural or comparative approaches.

"Interculturality" can be regarded as a subject for research only if the interaction process of the relationship of A to B is considered from a *generative* perspective that goes beyond fixed points based on content (Culture "A," Culture "B"), to address the "area in-between" the two cultures. This "area-in-between" or the "interculturality" [7, p. 345] is a process whose essential characteristic is that culturally determined behavior patterns of both "A and "B" change, through which a new regenerating quality "C" arises. "Interculturality" is therefore not synthetic in nature, but synergetic.

THE LIFE-WORLD MODEL

If we are to understand intercultural communication from an interactive perspective as interpersonal, communicative action between differently enculturated individuals, then it becomes clear that the two preceding models must be combined. Their form, however, must be differentiated. The importance of communicators' expectations of differences and their meta-images in intercultural communication makes it necessary to analyze one's own as well as the other's

[2] G. Hofstede surveyed a total of 117,000 IBM employees in sixty-six countries concerning personal values and goals, perception of the work situation, and job satisfaction, thus producing the most comprehensive empirical study to date for comparative management research [3, 4].

[3] See overview of A. Thomas and K. Hagemann [6] in N. Bergemann and A. L. J. Sourisseaux [pp. 173-199].

possible expectations and their consequences for the action that takes place. The first task can be achieved by reflecting on (1) one's own and (2) the other's (and in this sense *intra*cultural) enculturation process. That is to say, knowledge of one's own as well as the other's enculturation process is transmitted both in synchronic and diachronic form. An analysis of the consequences for a negotiation process can be carried out to a limited extent 3) through the documentation and simulation of examples of interactions. However, equally important in this context is a sensitizing, non-specific cultural training approach that enhances the ability to empathize in such intercultural situations.

The implications of both the culture-specific and the culture-sensitizing approaches can be observed in the "life-world Model" as originally formulated by Schutz and Luckmann [7] and which J. Habermas [8] modified and integrated into his communication theory.[4] Both models represent, *mutatis mutandis,* a concept in which "life-world" is understood to mean the everyday self-reproducing area of one's life. Although the question of interculturality is not explicitly dealt with in either of the approaches, it can be derived if we apply the formal analysis of the "life-world" to empirical data, thereby automatically arriving at distinguishable fields of communicative action. In reality, of course, the areas overlap and in this respect Figure 1 below showing intercultural process is somewhat over-simplified. It does show, however, to what extent intercultural interaction, is a unique, non-repeatable event. As such, it must be regarded as a constant negotiation with the momentarily present "inter-culture" and therefore can be understood as "interplay."

Our point of departure here is to look at the formal analysis of the structures in the "life-world" model. Habermas distinguishes between the three structural components: culture, society, and personality:

> For me *Culture* is the store of knowledge which supplies participants in communication with interpretations when they are communicating about something in a specific "world." *Society* is for me the legitimate order by which participants in communication regulate their membership to social groups, thereby ensuring a certain solidarity. *Personality* I understand to mean the competencies which enable a subject to speak and act and therefore enable him to take part in the processes of communication and understanding, thereby confirming his identity [. . .] The interactions that are tied up in a web of everyday communicative practice form the medium by which culture, society and individual are reproduced [8, pp. 209-210].

The interdependence of culture, society, and personality is clear according to a theory of action. For example, an individual's actions are never without extra-mundane aspects, but take place against the background of a socially mediated, "cultural" store of knowledge. The reverse is also true: these actions generate those systems of signs that constitute the everyday practice of social groups, and

[4] For Habermas' critique of Schutz and Luckmann see [8, pp. 194-195].

they can become embedded in the store of knowledge and/or generate it. Accordingly, sign systems are "represented structures, which are formed intersubjectively, deposited historically and mediated socially" [8, pp. 208-209]. This means that the life-world with reference to "the life-world in its totality" is the "arena as well as the destination of my and our reciprocal actions. In order to reach our goals we must come to terms with the facts as they are and change them. We act and therefore are effective not only within our life-world but on the way to a different one" [8, pp. 28-29].

In looking at Schutz and Luckmann's idea of the "scope" [7, pp. 63-64] of the life-world, we can assume that, despite a certain degree of overlap, different social groups possess different stores of knowledge. It therefore seems appropriate, given the necessity to define meanings clearly, that we speak of communicative interaction between different life-worlds. This could, for example, refer to contacts between French and Russians. It could, however, also by definition refer to *intra*cultural interaction between Californians and New Yorkers. The dividing line between life-worlds is in any case a heuristic one. As a result, Figure 1 cannot be empirically verified in this form, since, as already mentioned, individual life-world areas are bound to at least partially overlap. It does, however, illustrate the interaction process.

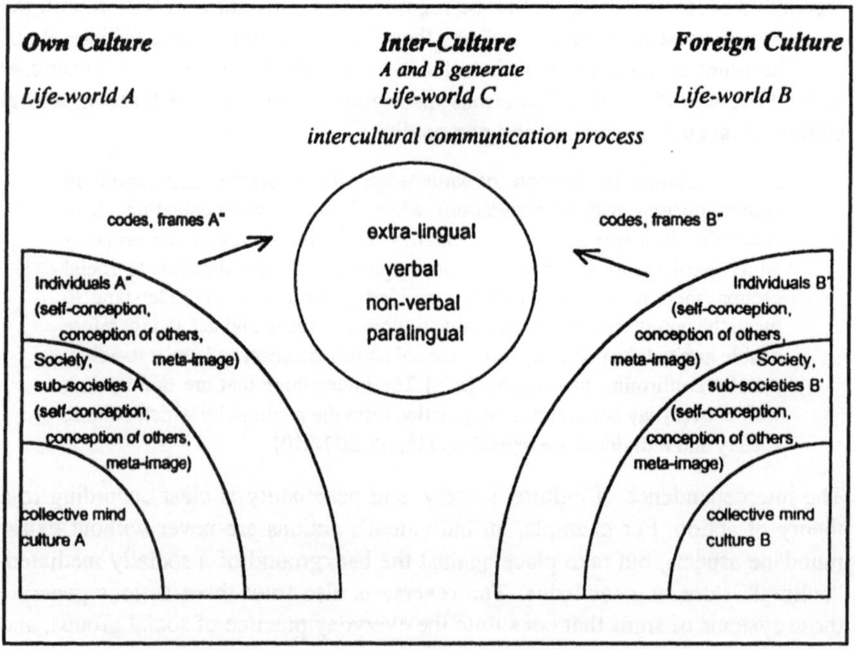

Figure 1.

As Figure 1 shows, interculturality can be understood as a process, as an event that results from an encounter between two or more people, which is specific to culture and individual and is, from a hermeneutic point of view, unique. The event itself, i.e., the specific interaction, occurs within and as the "interplay" between the life-worlds. The interplay itself can for various reasons neither be determined nor classified *a priori:* 1) Direct interactions are always dependent on the individual. Individuals' images of themselves, of the "other" and of what they imagine the other individual's view of them to be (meta-image) are however not the same and can vary dramatically according to age, socialization, knowledge of the world, knowledge of the "other" life-world, empathy, etc. This means, for example, that a certain German businessman does not have a meeting with "a typical Englishman" (i.e., one whose behavior is predictable), but with a certain Englishman, who in turn has a very specific and individual self-image, image of the "other," and meta-image, and 2) The interaction situation itself can, as far as the course of events is concerned, be likened to a constant negotiation of communication and behavior strategies to produce consensus. Important here is the bilateral relationship between readiness to assimilate and the demand on others to assimilate. To what extent and in what way one adapts to one's partner can more or less be determined only in the situation itself and will be different every time.

From an individual-psychological perspective, it would seem reasonable to work on a "culture-sensitizing" basis rather than a culture-specific one. The latter is more relevant on the border of social psychology and the social sciences. More specifically, a culture-specific framework is especially valuable for answering questions about the extent to which an individual's actions are determined by socially conditioned schemata of perception and experience: "Cultural values shared by one group may be rejected by another. The values are learned by the members of the group, and hence taught by other members. A culture is passed down from one generation to the next. It is acquired and is not innate" [9, pp. 14-15]. R. Munch, taking as an example the media coverage of the democratic movements in Eastern Europe at the end of the eighties, convincingly demonstrated the connection between advances in communication technology and changes in society's values. His thesis that "the revolutions in Eastern Europe in the Autumn of 1989 [. . .] were the first in history in which telecommunication played a decisive role" [10, pp. 113-114] is surely correct. It confirms at the same time that the acquisition of culture-specific knowledge will in the future increasingly have to take into account the fast moving and in this sense *intra*cultural dynamism of foreign life-worlds. As the individual's "store of knowledge about his life-world [. . .] is largely taken from experience within the group" [7, pp. 30-31], it is necessary for intercultural training courses to adopt an approach that sensitizes as well as conveys knowledge about the foreign life-world. This knowledge must be conveyed both synchronically as well as diachronically. "Ready-made" and therefore static images of a society will in the future be less and less relevant for a reality that is changing at an ever-increasing rate and that is, in this respect, increasingly likely to lead to misunderstandings in the sense of *critical incidents*. It is therefore important to document life-worlds in a historical context and thus illustrate their intercultural processes. This relates to the "other" as well as to the "self," as it is only in this way that the interdependence of self-image, image of the "other," and meta-image, as well as the effects of anticipating differences, can become transparent. This underscores the necessity to move beyond the statistical comparisons that often mark "regional/cultural studies."

Despite the rapid changes in society's values and norms, it can be assumed that parts of the individual's store of knowledge are based on latent thought and behavior patterns that are resistant to social change. J. Assmann distinguishes here between *cultural* and *communicative memory:*

> We understand the term "communicative memory" to include those parts of the collective memory which are exclusively devoted to everyday communication [. . .] All studies seem to confirm that it normally stretches back no more than 80 to (at the very most) 100 years [11, pp. 10-11]. In contrast to this, "the cultural memory" is characterized by its remoteness from everyday life. This remoteness (transcendence of everyday life) characterizes its time horizon. The cultural memory has fixed points and its horizon does not change with the ever moving present moment. These fixed points are important events of the past the memories of which are kept alive through cultural formations (texts, rites, memorials) and institutionalized communication (recitation, observation, contemplation) [11, pp. 12-13].

Assmann is correct to insist that this store of knowledge differs in content not only from culture to culture but also from epoch to epoch. The relevance of this knowledge for today's society is thus put into question. Moreover, not all aspects of a society's cultural memory carry the same relevance in every period. Therefore, focusing on supposed universalities within the interacting life-worlds would seem to be of little value. Studies in the area of intercultural communication should rather concentrate more intensively on "seismographically" illustrating the interdependence of the three (heuristic) horizontal levels of a given life-world at a specific point in time.

The following example of an intercultural consulting and training seminar will serve to demonstrate how this concept of intercultural management research can be applied in practice.

INTERCULTURAL MANAGEMENT CONSULTING: THEORETICAL PRINCIPLES AND CASE STUDIES

One of the most important problems of intercultural management research is that it is even more interdisciplinary than contrastive approaches and that it is under a much more direct pressure to produce practical applications than the individual disciplines. This becomes clear when one looks at the spectrum of activities associated with intercultural management consulting. It should be mentioned that the following list only represents the most common consulting areas.[5]

- Personnel selection for assignments abroad; development of intercultural assessment centers; reintegration consultation
- Consultation to determine limits for standardization, e.g., in intercultural marketing; description and analysis of cultural specifics of oral and written communication systems
- Corporate communication conceptions in multinational companies or joint ventures
- Description of causes and changes in cultural values in the company's target markets
- Consultation for internationalization decisions and processes, including the analysis of potential for misunderstanding as well as synergy
- Communication analysis of specific interaction processes within internationalization.

Case Classification and Description

Each consultation is generally linked to an actual case that requires a problem-solving strategy. In order to be able to operationalize this, it is necessary 1) to classify and 2) to describe the case in as much detail as possible.

1) There are two categories of case classification. The first instance is where the interactive context is already "intercultural." In these cases, the parties involved are aware of the internationalization process, and the emphasis is on finding possible causes for dysfunctional occurrences, such as common "critical incidents" or the threatened failure of an international project, from which a problem-solving strategy is developed. It is also conceivable that existing interactive contexts could be analyzed with regard to their optimalization (e.g., their potential for intercultural synergy).

The second category comprises situations when the interactive context does not yet exist, and the internationalization project is only in the planning stages. In these cases, the participants of the future intercultural interaction are both known in terms of their respective domestic cultures. The context for such interactions cannot be analyzed deductively, only inductively inferred or anticipated. Thus, the key is to obtain as much reliable evidence as possible as to how the planned interaction processes could evolve and where, for example, the

[5] The areas of activity correspond to the concentration areas of the M.A. program "Intercultural Business Communication" at Friedrich-Schiller-Universitat Jena/Germany.

interaction partner's limits of acceptance[6] can be expected. In practice, these questions arise when companies plan international cooperations or "go international" in marketing terms.

2) Depending on the classification, the case description should be oriented toward the following perspectives and should outline the current scenario in as much detail as possible: Who are the (potential) interaction partners and which goals are they pursuing? Have any previous interactions taken place between the partners? If so, what was the outcome?

In all cases mentioned, it is essential to formulate precisely the reasons for the consultation and the goals to be pursued.

Preconditions for the Partners' Actions

Whether an interactive context exists or not, the second step of the consultation is to carry out a far-reaching investigation of the possible life-world actions of the (potential) interaction partner. Specifically, the consultant seeks to uncover why the partners act the way they do in their *own* culture in specific contexts.

Much of the question is linked to research in those aspects of cultural and social anthropology, ethnomethodology, social theory of action, and sociology of knowledge, which deal with structuring aspects of everyday life actions.[7] In general, the issues at the heart of the discussion are 1) structural characteristics and 2) functionality of everyday thinking.[8] A well-conceived consultation will constantly return to this idea that a *specific* context must remain as a frame of reference.

Knowledge of Everyday Life

Knowledge of everyday life is defined according to Patzelt as "the result of every biographically specific experience of the everyday world . . . that a person has at a given point in his or her life" [13, p. 45]. Knowledge of everyday life thus describes "all stores of knowledge and interpretive routines of a person which serve to implement his or her ethnospecific everyday life experience" [13, p. 45]. It is first handed down historically through the "cultural memory"; second, conveyed socially; and third, acquired through one's own experiences. These three levels of cultural, social, and individual knowledge are interdependent and form a stock of interpretation upon which the life-world context is interpreted and constructed. This stock of interpretation itself exists potentially as an "archive," i.e., it is never activated in its entirety. It always functions solely as a subsystem that is

[6] See J. Bolten [12].

[7] "Everyday World" (Alltagswelt) is interpreted here in accordance with Patzelt's conception of a social field "in which the everyday reality is experienced and maintained" [13, p. 43].

[8] I refer to Patzelt's heuristic distinction between "everyday knowledge" and "everyday thinking" [13, pp. 45-46].

assigned practical relevance in a given situation by everyday thinking which ensures normality by assigning meaning to actions.[9] Thus, everyday knowledge remains unanalyzed to a large extent. It becomes the object of reflection only in situations that appear implausible and unclear. Even in such situations, however, everyday knowledge cannot be explicitly and unambiguously formulated, because it sets up specific contexts based on its individual and social distribution.

Everyday knowledge can also be described using the ethnomethodological concept of indexicality.[10] According to this concept, stores of knowledge are fundamentally ambiguous, and meanings depend on the context or subsystems in which they are activated. Several assumptions are necessary to make interactions plausible and to bring interpretation into harmony, or deindexicalized. First, we must assume that the components of everyday knowledge constantly express more than meets the eye. Second, we must assume that interactions are characterized by common subsets that have various functions. In specific situations, these subsets must reduce complexity by means of an "automatic" deindexicalization mechanism, and they must ensure intersubjectivity. The probability of encountering another with whom we share such a common subset in a routine situation decreases, the more personal the experience through which a store of knowledge was conveyed. Conversely, the probability increases the more the stores of knowledge derive from cultural memory. In this respect, cultural memory represents that last common interpretive basis for trouble-free everyday interaction when the subsets of individual or social knowledge together are not enough to bring about a sense of plausibility. At the same time, this means that members of an ethnic group define themselves on the basis of this cultural store of knowledge.[11] The less that individuals can draw upon a common cultural store of knowledge, the more "foreign" the situation appears. The concept "interculturality" and "interculturality" can also be distinguished in this context: intraculturality describes situations in which participants share a common cultural store of knowledge, whereas interculturality presupposes at least a partial inequality.[12]

[9] "In this context, one can also say that everyday knowledge is a large store of 'unpublicized knowledge' that lies under the layer of intersubjective 'public' communication" [13, p. 47].

[10] See [14].

[11] Patzelt commented on this: "people who possess such a large amount of shared common knowledge from the multitude of possible stores of everyday knowledge that they can routinely and satisfactorily carry out consistent reality-constructing coordination of interpretation of meaning and actions are called 'members' of that ethnic group; these can be distinguished from the shared ethnotheories of other ethnic groups" [13, p. 47].

[12] The transition is of course fluid as could be shown with border situations. An example is of the relations between East and West Germans: Developments that took place over forty years of separation have surely led to a considerable incompatibility of individual and social stores of knowledge; and, due to the interdependence of the three levels, have had effects on the cultural store of knowledge which is considerably resistant and is passed on over several chains of generations.

Everyday Thinking

While everyday knowledge is fixed in terms of content, the concept of everyday thinking refers to *functional* aspects. It concerns the issue of *how* the store of knowledge is dealt with and which procedural rules are used to perform meaningful everyday actions. The energy principle, as formulated in the disciplines of cultural and social anthropology, plays an essential role in this relationship. It states that the goal of each act in any given situation is to reach the maximum benefit with a minimal expenditure of energy.[13] When applied to the functioning of everyday thinking, this means that during deindexicalization, for example, the tendency will be to privilege those strategies that require the least energy expenditure and that lead to the greatest possible sense of plausibility. The methods of interpretation that are routinely used by everyday thinking follow this rule. It is always a matter of decoding and encoding meaning in the interpretive framework of goal-means-relationship, or of recognizing causal conditions that do not infringe on the expectations of the other and the self for normality. Even if such infringements do arise, everyday thinking is able to functionalize stores of knowledge that are accepted as problem-solving strategies in an interpersonal, social, or cultural way. Among these are explicitness (e.g., certain forms of meta-communication), avoidance strategies (e.g., leveling off, extraterritorialization), or repair mechanisms (e.g., apologies). Implementing these strategies in the interaction depends on the context and follows the energy principle: everyday thinking—whether context-generated or context-generating—combines the stores of knowledge so as to maximize the probability of expected outcomes. Routine actions are the result of such thinking. On the one hand, they partially fossilize certain combinations of stored knowledge as independent subsystems. On the other hand, the systems remain nonetheless open because, as already shown, the store of knowledge exists as a potential archive and thus theoretically allows for any number of other combinations. Practically speaking, this is the case if the interpretation or construction of meaning in a partial context loses plausibility because the overriding context has changed (by means of the systemic connection). In light of this, other combinations become more plausible and expectation of their probability increases, so that competing, but possibly also dominating, systems of interpretation arise. A typical example of this are the processes involved in situations where values change. Everyday thinking, like everyday knowledge, is purely context-bound and can thus be documented by analyzing specific actions. This is confirmed for consultants by the high degree of precision

[13]A somewhat trivial example to illustrate the point: there are two apples of the same quality hanging on a tree. One can be reached without a ladder, one can only be reached with a ladder. If the goal is only to pick an apple, nobody would think of making the effort for the one which is harder to reach. If it is larger or better in quality, the increase in quality (e.g., in calories) and the necessary increase in expenditure of energy to obtain the apple will be put into relation. It is worth the effort to climb the ladder if the increase in energy input is under the level of energy output.

required to describe completed or sought after interactive contexts, as described in the first section.

There is no such thing as *the* behavior of *the* members of the ethnic group X. Even under the premise of a common cultural store of knowledge, the spectrum of meaning is open to any number of associations depending on the subsystems and individual knowledge (which are again open to new interpretation). For this reason, it is necessary during the analysis of culture-specific conditions to obtain exact information about how persons interact in a comparable intercultural situation or about the way stores of knowledge are drawn upon during the intercultural construction of reality.

The problem is that, together with the stores of knowledge, intracultural expectations of normality are not sufficiently explicit to be described in terms of a fixed body of knowledge; they can be explored only by means of a situation that infringes on these expectations.

Exploring the infringement of expectations nevertheless makes a cultural comparison possible, provided that the same context is selected for the intercultural analysis. Among other things, this comparison can give information regarding whether infringements on expectations of normality in culture A can also be registered as such in culture B, and to what extent problem-solving strategies in the two cultures are compatible.

Intercultural Interaction

As already mentioned, studies based on cultural comparisons do not provide information or permit predictions about the process of intercultural interaction because cultural interactions generate a "third" life-world. This "third" life-world can possess different conventions for action in contrast to the intracultural contexts of the respective partners. Based on this, it is impossible to predict or plan intercultural interaction in detail. In principle, one can assume that the basic rules for human behavior agree in structure with those used in intercultural interaction. The energy principle is of utmost importance here, in the sense that one tries to expend a minimum of energy in order to obtain a maximum of meaning construction, deindexicalization efficiency, plausibility, or "normality." Analogous to the example of the apple harvest (see footnote 13), this means that the *relative* minimum level of energy will be applied (which may well be considerable in absolute terms). Measuring this relativity depends on which variables constitute the interactive context. The following variables have a prime importance in this area:

1. the duration of the interactive relationship,
2. the subjectively perceived degree of alienation in terms of the interaction context, and
3. the degree of self-consciousness with which the participants interact.

The duration of the interactive relationship is proportional to the degree of familiarity of the participants, when referring to primary experiences.[14] Obviously, the longer a specific "interculturality" exists, the greater the degree of familiarity the participants will have. This correlates with the subjectively perceived degree of alienation and the degree of self-consciousness with which the participants interact. Generally, the degree of alienation and self-consciousness decreases with increased familiarity. In terms of structure, this process parallels analogous instances of intracultural reality construction; but, in terms of content, there are two essential differences. First, the subjectively perceived degree of alienation and self-consciousness initially are relatively higher. This includes a conscious anticipation of the other's expectations, as well as the introduction of deindexicalization strategies that attempt to eliminate the intercultural dimensions of everyday experiences.[15] Second, as already mentioned, a common cultural store of knowledge, which secures plausibility to the highest degree, can only be called upon to a limited extent, if at all.

In accordance with the energy principle, everyday thought attempts to impose its assumptions of normality on the intercultural sphere as quickly as possible. This generally seems to succeed in international business because of pervasive globalization processes. Space, time, and distance have become almost insignificant; external differences (e.g., clothing, furnishings in conference rooms or hotels) are seldom manifest on an international level, and behavioral conventions in negotiations follow a more and more generally accepted "world standard." Despite this, one should not forget that interpretive processes call upon differing cultural stores of knowledge. Thus, on the surface, deindexicalization processes may seem relatively uncomplicated because individual experiences or the professional socialization process in multinational companies combine to form subsets. However, compatible subsets on the surface do not necessarily correspond with compatible deep structures, so that apparently similar stores of knowledge can be put together in completely different interpretive systems without the participants becoming aware of it. Since a similar phenomenon also characterizes the implementation of problem-solving strategies, actual indexicalized situations tend to be reinterpreted as deindexicalized ones in order to maintain a sense of plausibility.

As a result, participants subjectively perceive less and less alienation the longer the duration of the intercultural contact; correspondingly they also become less self-conscious about their actions. The objective basis of understanding, however, becomes narrower and narrower because the interpretation of meaning

[14]In terms of secondary experiences, they have been passed on by means of images, or possibly stereotypes and prejudices, and therefore a high degree of familiarity can already exist of course before the beginning of the interaction. This especially influences the first phase of the interaction.

[15]There are exceptions, such as in crisis communications.

is defined more and more by the premises of one's own culture. If one becomes conscious of deindexicalization problems as such, they carry less weight, because the previous construction of meaning did not take place interculturally and thus does not possess any consensus-forming function. Points of view of behavior patterns of the partner from another culture can no longer be accepted because they are viewed as infringing upon one's identity or upon the culturally conventionalized limits of one's tolerance. This is also a reason why international business ventures frequently encounter crisis situations only after a fairly long period of apparently problem-free interaction. Although critical incident analyses are useful principally for already existing interactive relationships, the main issue in planning for internationalization is rather how the potential for synergy can be determined. Synergy is present if the effectiveness of an (intercultural) interaction is greater than the sum of the (in each case intracultural) effects of the participating individuals.[16]

INTERCULTURAL MANAGEMENT TRAINING: A MODEL

The explanations in the first two sections of this chapter suggest objectives for an intercultural management training seminar. To generalize, the aim of the training program is to empower international management personnel to act competently in targeted cultures. In addition to having sound synchronic and diachronic knowledge of the target culture, intercultural competence requires fundamental abilities and skills. Individuals should be able to:

- formulate self-conceptions, expectations (image of the other), and one's own anticipation of the other's expectations (meta-images) with reference to specific interactive contexts and be able to recognize their interdependency as the basis for one's own actions
- perceive and describe changes in their own behavior in intercultural contexts relative to communicative contexts in their own culture
- "endure" deficits in plausibility and experience insecurity in intercultural contexts in the sense of a tolerance for ambiguity
- negotiate and implement conventions in action and communication in intercultural contexts
- identify the potential for both misunderstanding and synergy in specific intercultural situations
- recognize communication styles and problem-solving strategies in terms of their cultural basis

[16]See Ropella [15, pp. 174-175].

- maintain role distance in terms of one's own communicative behavior and empathy for the other
- follow multilingual negotiations
- apply metacommunicative strategies.

The overwhelming majority of intercultural training seminars carried out today worldwide follow American training models. Examples include cognitively oriented exercise forms such as the culture assimilator, culture capsules, etc., or they are based on intercultural sensitivity oriented games like Bafa-Bafa or Bargna. Without going into in-depth critiques for each of the models at this point,[17] one can identify an essential shortcoming of almost all of these conceptions, namely that all of them miss their goal by failing to create a realistic intercultural context. At best, cognitively oriented training programs enable a reflection about the intercultural context from the observer's point of view. Although empathy-promoting games and simulations provide experience with situations involving alienation, they always run the risk of not being taken seriously by the participants due to their extremely fictitious contexts for action ("you are now a member of the 'Adamas' culture and must now negotiate with representatives of the 'Omera' culture").

From the perspective of an interactive theory, it is necessary for international management assignments to create intercultural contexts that require participants to act in ways similar to those that would be required of them in real situations. In terms of methodology, simulations are most suitable for this purpose because they are based on case studies from the international business world and follow the goals outlined above. The following outline shows how such a simulation can be conceived using the example of a training seminar created and tested in the Department of Intercultural Business Communication at the University of Jena, Germany.

Target Group

The training seminar is geared to management and management trainees as well as business students who are preparing for assignments in international management. The requirements are very good language skills in the *seminar language* (German) and in at least one other *simulation language* (English, French, or Russian). The seminar group should have at least eight but no more than sixteen people. Depending on the country of origin, the participants are put together multinationally. Ideally, there should be at least two participants in each team from France, Great Britain, Russia, and the United States. If this is not feasible, a trainer from one of these countries can also act as a participant.

[17] See Thomas and Hagemann [16] and B.-D. Muller [14].

Seminar Leaders

In the ideal case, the seminar is led by a native German speaking trainer and four native speaking co-trainers for American English, British English, French, and Russian. The minimum requirements for a successful training seminar would be one trainer and two co-trainers with differing native languages.

Technical Equipment

To implement the training seminar, four small offices (with telephones, if possible) and two large conference rooms are generally required. If these facilities are not available, two small rooms as well as two larger rooms would be the minimum requirements. Additionally, two camcorders, one overhead projector, and four to five calculators should be available.

Training Materials and Framework of the Simulation

The seminar switches between plenary and group work. A total of four groups are formed, each representing a company from France, the United Kingdom, Russia, and the United States. The groups reach decisions for their "company," discuss possibilities for joint ventures and carry out negotiations to reach these goals. The corporate language is the language of the location of the company. With cooperative and joint ventures, the language of negotiation is determined by the participants themselves. The plenary meetings, which take place in German, provide for evaluating videotapes of individual groups, as well as for instructing interculturally relevant strategies for negotiation and behavior.

The participants receive training materials for their respective target cultures, as illustrated in Figure 2. The case description and starting situation of the company they will be representing during the simulation should be thoroughly read by the participants. The training materials contain the texts for all of the assignments and information that the participants require for the exercises in the plenary meetings and for decision-making in the companies (in individual groups). The updated market reports are not initially provided. The market reports are provided in the trainer handbook and are handed out to the groups before the beginning of each round in the simulation. One can expect six hours for the group work and six hours for the plenary work. A total of 1.5 or 2 days should be available for the training seminar.

CONCLUSION

As training seminars with international student groups and in multinational companies have shown, the participants very quickly forget that the seminar is a simulation. This is very important because participants will only experience the

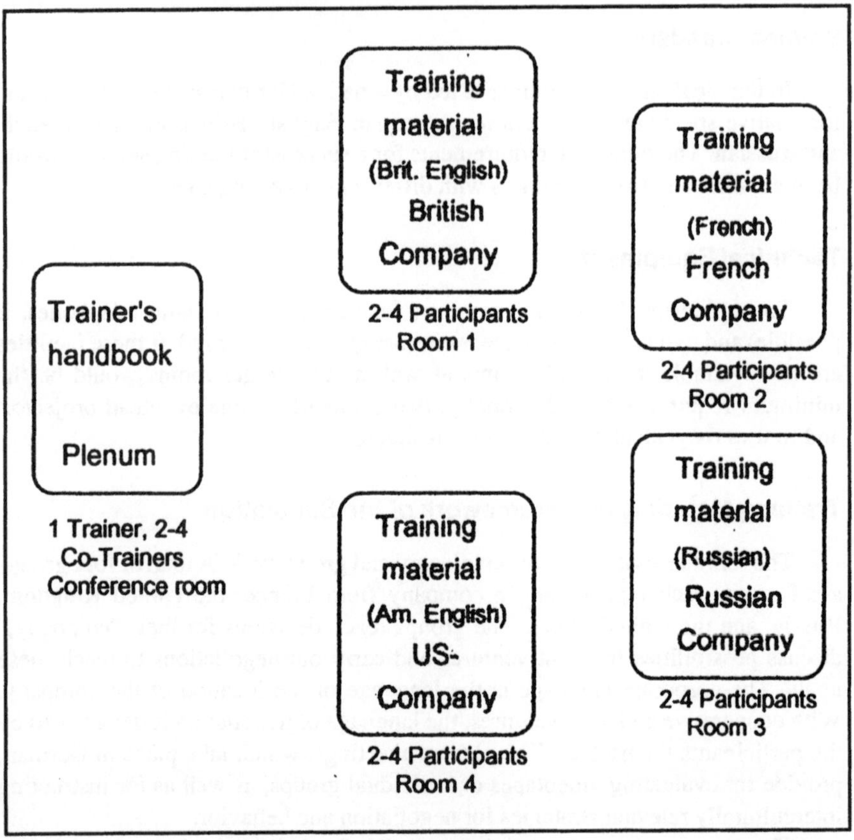

Figure 2.

training seminar as an authentic situation if it initiates an intercultural process that approximates realistic conditions.

Hence, the success of the simulation depends on the degree of authenticity: the more the participants can identify with the context of the simulation, the more authentically they will react in the intercultural exercise sequences. The analysis of the videotaped sessions then reveals very clearly how the intercultural competence of each participant can be improved.

Intercultural competence is a general ability. Therefore, the specific field of application (i.e., business) is not that important for the choice of what is taught; this ought to be oriented to the objectives described at the beginning of this section. This is just as true for a corporate training model as for a classroom teaching model.

REFERENCES

1. J. Hentze, Kulturvergleichende Managementforschung, *Die Unternehmung, 41,* H.3, pp. 171-185, 1987.
2. N. Kumar, Interkulturelle Managementforschung, *Wirtschaftswissenschaftliches Studium, 17,* H.8, pp. 389-394, 1988.
3. G. Hofstede, *Culture's Consequences,* Sage, Beverly Hills, California, 1980.
4. G. Hofstede, *Interkulturelle Zusammenarbeit. Kulturen—Organisationen—Management,* Aus dem Engl. von Nadia Hasenkamp und Anthony Lee, Wiesbaden, 1993.
5. K. Knapp and A. Knapp-Potthoff, Interkulturelle Kommunikation, *Zeitschrift fur Fremdsprachenforschung,* H.1, pp. 62-93, 1990.
6. A. Thomas and K. Hagemann, *Training interkultureller Kompetenz,* in N. Bergemann and A. L. J. Sourisseaux, pp. 174-200, 1992.
7. A. Schutz and T. Luckmann, *Strukturen der Lebenswelt,* 2 Bde., Frankfurt/Main4, 1991.
8. J. Habermas, *Theorie des kommunikativen Handelns,* 2 Bande, Frankfurt/Main, 1981.
9. G. H. Mead, *Mind, Self and Society,* C. W. Morris (ed.), University of Chicago Press, Chicago, 1934.
10. R. Munch, *Dialektik der Kommunikationsgesellschaft,* Frankfort/Main, 1991.
11. J. Assmann and T. Holscher (Hg.), *Kultur und Gedachtnis,* Frankfurt/Main, 1988.
12. J. Bolten, Life-World Games, *European Journal of Education, 28,* H.3, pp. 339-348, 1993.
13. W. J. Patzelt, *Grundlagen der Ethnomethodologie,* Munchen, 1987.
14. B.-D. Muller, Grundpositionen einer interkulturellen Didaktik des Deutschen als Fremdsprache, in *Praludein, Kanadisch-deutsche Dialoge,* B. Krause (Hg.), pp. S.133-156, Munchen, 1992.
15. W. Ropella, *Synergie als strategisches Ziel der Unternehmung,* Berlin, New York, 1989.

CHAPTER 7

A Relational Framework for Professional Communication in International Organizations

FREDERICK STEIER

The rapid globalization of business and professional organizations has heightened a need for, and given rise to, new forms of intercultural relationships. These have afforded new communicative opportunities. Yet, these increased intercultural opportunities have not necessarily led to improvements in mutual understanding within and across international organizations. Indeed, the problems created have gained more attention than the opportunities offered. A recurring question for practitioners trying to facilitate cooperative work across cultures is how best to think about (and act with) cultural differences. How, for example, might we move beyond seeing differences as a set of fixed contrasts which may or may not be "relevant" to our intercultural partners and to ourselves, to seeing differences as jointly constructed and interactionally situated? In this chapter, a framework is developed for conceptualizing cultural difference as a mutually appreciative system—a strongly relational framework. However, rather than stop at the point of grasping the emergence and salience of cross-cultural differences, we must understand how people who populate these systems of difference might actually work together—surely a question of great import to communication professionals. Thus, this chapter seeks to move from the recognition of cross-cultural differences to the development of a framework for *inter*cultural communication. This framework must be a relational one. Perhaps it is best to begin with an example.

INTERCULTURAL ISSUES AT GOTHENBURG

Some months ago, Hans, the Chief Financial Officer of the American office of a large, international Swedish manufacturing organization (we can call it the Gothenburg Corporation), told me about a dilemma in his organization that he said had to do with professional communication. The dilemma arose, he said, at a recent marketing meeting where the Gothenburg Director of Marketing for American operations (also Swedish—let's call him Gunnar) was meeting with American regional marketing directors. One of the purposes of the meeting was to set sales quotas for a sales period. During the meeting, Gunnar mentioned that he was interested in seeing sales increased. Specifically, he "stated" (let's get back to the difficulty of choosing the correct word here shortly) to the American director for region C, Tom, that sales for the specified period should be 900 units. Tom, according to Hans, the CFO, looked a bit nonplussed, nodded, but didn't say anything.

In the days following, two dramatically different versions of what had taken place at that sales meeting emerged in conversations between Hans and each of the participants. Gunnar's account centered on themes of surprise and disappointment. He wondered why Tom hadn't responded to his offer. He knew that 900 units was unusually high as a sales quota, and would have been a record by quite a large amount for the period in question. He felt, however, that it was a nice first step as an "invitation" to have a dialogue about what a reasonable but aggressive sales figure might be. Tom could have responded with a lower figure, and they could have had the quota emerge jointly between them. This was the way the process should be between the national and regional directors. Tom just seemed to accept the figure, yet Gunnar knew that the figure couldn't possibly be met. Didn't Americans understand what participatory management, which was so well understood in Sweden, was about?

Tom's versions of the conversation were rather different, and Hans wondered, perhaps only half in jest, if Tom and Gunnar had really been at the same meeting. Tom, who had a long history of technical sales work, and had extensive work with the military, couldn't believe that Gunnar had "commanded" him to meet such a high sales quota. He was used to carrying out director's wishes, but this really seemed excessive to him. Perhaps, he suggested, things were different in Sweden, and the directors were supposed to be more dictatorial and/or unreasonable. He complained bitterly (but not to Gunnar) about what he felt was an unreasonable request.

We might observe that the dilemma was rooted in cross-cultural differences in understanding that initial "request" (or was it a statement, or a command, or an invitation?). From Tom's point of view, it was a command, for that is what directors do, and should be carried out. From Gunnar's point of view, it was an invitation, and should be followed up by some form of negotiation. It was not a question of who was right. Had Tom and Gunnar been familiar with speech act theory, which offers a way of thinking about "what we do with words," they might have been able to recognize that "the same utterance" could be offered or heard as a request or an invitation [1, 2]. But would they then have been able to work together to resolve their discrepancy?

From Hans' point of view, each was correct within his own frame. An awareness *that* what is said can be listened to differently from within a different frame is a first step toward collaborative work in international organizations such as Gothenburg. What is also required is an awareness of how different interpretations can reasonably be constructed, and *what* these interpretations might be. Rather than label something as a "misinterpretation," what if we open ourselves up to the possibility that what we have are multiple interpretations, where even the speaker is not privileged to *the* correct interpretation of her or his own utterance (nor is the listener to her or his hearing). Hans could see that both Tom's and Gunnar's interpretation of the situation made sense based upon their own history of being involved in similar situations and interactions. This history clearly was tied to (although not necessarily determined by) their cultural background, with the Swedish orientation being more collectivistic than the highly individualistic American orientation [3]. What Hans wanted to know, however, was how to create a communication framework where Tom and Gunnar would work together. To Hans, the challenge clearly centered on professional communication.

Hans and I were on a long training run together when he had posed the problem, so there was plenty of time to reflect on the issue, plenty of spaces between our conversation, as we worked our way together along the trail in the woods. Sometime later, Hans continued by observing that it was not just a matter of awareness. If, for example, Gunnar had had a training seminar dealing with American cultural patterns in organizations, Gunnar might have expected Tom to interpret his utterance as a command, and acted accordingly. If, at the same time, Tom had had a similar training seminar dealing with Swedish cultural patterns, it is possible that he might have heard as an "invitation" what Gunnar (in trying to be more "American") now intended as a command. They could still be operating out of different frames, although now out of an awareness of the other's frame. The potential permutations are dizzying, but the problem of frames for collaborative work would still be there, training programs or no training programs.

Solutions to these dilemmas are not always easy, and my intent here is not to offer an example of an organizational communication consultant's success story. Rather, what I would like to do here is to develop themes that emerge in grappling with this situation from a framework that makes sense for professional communication in international organizations. This framework presupposes that communication processes are rooted in human relationships. While this simple statement might not be particularly remarkable to most readers, it may well be that we haven't to this point taken the idea of a relational framework seriously enough to allow it to show up effectively in our everyday practice. Perhaps it is due to epistemological assumptions that make it difficult to appreciate, or even render visible, the connectedness of the living systems that make up our environments (what Capra calls "The Web of Life" [4]), and our social worlds. That is, a lack of appreciation of interconnectedness in our "natural world" is also revealed in how we construct our worlds in social settings, such as organizations, in relationship with others. In the next two sections, building upon the above "story," ways of approaching understanding in international organizations that begin to question our usual assumptions are developed.

INTER-PERSONAL NARRATIVES IN ORGANIZATIONAL CONTEXTS

As organizations (be they political, scientific, or commercial) have moved across international boundaries in their operations, there has been an increasing concern with what happens when the interactants from different cultures meet and try to work together. Clearly, there is a difference here between understanding cross-cultural differences and developing *inter*cultural understanding. The development of intercultural understanding is not only a theoretical concern for researchers, but also a challenge to practitioners. It is at this juncture between theory and practice that intercultural understanding becomes a critical concern for professional communication.

The dilemma in the Gothenburg organization highlights interpersonal communication in intercultural organizational settings. The dilemma in that story suggests several narrative lines. We have, just to name a few:

1. a narrative of how people from different cultural backgrounds hear differently and, as a result, act differently when presented with the "same" situation,
2. a narrative of how we need, in any international organization, to have knowledge of the cultural assumptions of the other,
3. a narrative of how we need to have knowledge of the frame within which the other (who may be from another culture) is operating. Notice how this is not the same as 2) above, since, as Hans noted in his conjectures regarding changing frames and training programs, people may always try to operate out of the "other's" frame, and
4. a narrative of the need to collaborate, to develop perhaps a mutual awareness of each other's ways of hearing and acting, which may even require the development of a "third language" [5]. This, in turn, requires that we hear and act in a more relational manner, understanding that everything we hear is heard in a relationship—a relationship of which we are a part.

Yet the story is also an organizational one, in that each of the above must be understood within the particular organization in which it occurred to appreciate the situation fully. It is here that the link to professional communication becomes even more marked, as we note the importance of appreciating our organizational contexts. In order to see the criticality of this context, it is crucial to recognize that in identifying and interpreting our choice of Gothenburg "narratives," we must make clear our epistemological assumptions about the worlds we inhabit when we work together.

FROM CROSS-CULTURAL TO INTERCULTURAL WORLDS

Intercultural encounters frequently pose problems for the realist framework. One version of this realist framework is that we live as observers of a *uni*verse that exists independently (of us and our knowing processes). What we claim as our knowledge reflects the reality of that independent universe. Within this realist framework, statements about the world we inhabit are supposed to be accurate representations of that world, independent of our language or "tools" for knowing. We can then, if we somehow have direct access to this independent universe, claim to be able to detect "mis-representations" of that universe.

Intercultural encounters often prove challenging within the realist framework, in that the others with whom we may be interacting may have what appear to us as misrepresentations, but which, from another perspective, can be viewed simply as different representations. Our worlds are different. Yet if we insist on holding to a realist position (that a correct representation is possible, independent of the cultural assumptions embedded in our language and ways of knowing), it follows that at least one of our different "worlds" is incorrect. We see this clearly in the way we label others' worlds as "naive" or "quaint" in their "mis"representations of the (or is it our?) world. If we assume that our own culture defines the framework for any intercultural encounter, we necessarily presume to have privileged access to *the* universe. We may, from a realist perspective, still feel a need to understand the others' world in order to work with them. However, the frame within which our attempts at collaborative work ensue is marked by an attitude of dominant knowing. For us, it is important to understand our others and where they are coming from, even if their knowledge, as such, is, to us, an illusion.

In intercultural encounters within a realist framework, the "interness" of intercultural communication is not realizable. Within this framework, we find competing "models" of reality, with any incommensurability resolved by what Bråten refers to as a "model monopoly" [6]. Bråten's model monopoly denotes a monolithic perspective that rules out other viewpoints not fitting precisely within that perspective, making a dialogic search for common ground not only impossible, but often undesirable. Most important to the intercultural encounters with which this chapter is concerned, this framework forces us to stop at the level of understanding of cross-cultural differences. There is no perceived need here to go beyond this understanding to question our own assumptions—including those embedded within our culture—to begin a path toward an intercultural perspective.

Yet, in stark contrast to the realist positions, there are epistemological positions that allow for, indeed, encourage, such a movement from *a* world of cross-cultural differences to worlds of intercultural understanding. These positions begin by questioning the very idea of a universe existing independently of observers. Within these positions, we question *how* we construct our worlds by critically examining what we do jointly as observers and actors. The social constructionist approaches [7, 8], and the radical constructivist approaches [9, 10], begin with the premise that our worlds are constructed, or even autonomously and jointly invented, by us, as inquirers who are simultaneously participants in those same worlds [11]. Rather than assuming the existence of stable realities, these approaches concern themselves with questions of how it is that we build up those stable realities that make up our everyday lives, and through which we coordinate our everyday activities. Despite differences between social constructionist and radical constructivist positions, both positions have in common the requirement of others in our worlds and the rejection of an independently (of us) existing universe.[1] Most importantly, constructionist approaches open the door to receiving others in intercultural encounters in ways that the realist positions do not. They focus, although in different ways, on human knowing as embedded in human relationships, allowing us to recognize that all we say or write has, as Bateson noted, *both* a report and command aspect [13]. (These report and command aspects have come to be more familiarly referred to as, respectively, content and relationship aspects of messages.) Constructionist approaches define an epistemological position that has critical implications for professional communication. In contrast, communication professionals have tended to focus exclusively on the content or report level, without a careful examination of what are the relational, or command, implications for those same activities. These have often been in contradiction with the report level itself.

As an example of this issue, in an attempt to create an electronic system for information and knowledge sharing at NASA (called the "Lessons Learned Information System") outside experts were brought to design the "architecture" of the system. This included not only software for the system itself, but also specifications for what entries to the system would need to look like to be considered usable. What was overlooked initially was the creation of an organizational culture that would support (and then, in turn, be supported by) the system of sharing information. Without the designers' and the resultant communication plan paying careful attention to the relational (command) aspect of messages about "how to use the system" (including training), these same messages about the system were heard by many potential users in a way that questioned the very

[1] For example, the starting point is our experienced world as a mental process (radical constructivism) experienced through processes of languaging, or it is our social realities constructed between us in interaction (social constructionism) [12].

sharing of information the system was designed to accomplish. Finally, in a redesign effort that questioned the assumptions embedded in the initial overall design, a shift in frames—from the system as a record of "facts," to the system as a communication medium inviting dialogue—was proposed [cf. 14, 15].

If taken seriously, constructionist positions regarding knowledge offer possibilities for conceiving of intercultural encounters as communicative opportunities, rather than merely as dilemmas. The following section proposes some "tools" to help guide that process. However, rather than thinking of tools as merely mechanical implements, I would like us to think of these as guides for action, for helping us orient ourselves in unfamiliar territory in a way that admits others into our intercultural worlds.

TOOLS FOR INTERCULTURAL WORK IN PROFESSIONAL COMMUNICATION

Part of the task of being a communication professional in international organizations is to help facilitate and develop an intercultural understanding in others in our organizations, whether in the design of documents, public relations campaigns, or training programs. As such, we need be aware of the frames through which others, even others in our own organizations, both hear and respond to us. What, then, are some tools to help guide this process? A set of candidates is offered below.

These "tools" represent ways for developing a *relational framework for communication in international organizations*—so necessary for moving from a perspective based on merely noting cross-cultural differences, to one based on intercultural understanding. Using these tools as a set can foster a commitment to being in a space with another in such a way as to afford truly collaborative work in international organizations.

Appreciative Systems

In his development of systems approaches to decision making and policy setting, Vickers became concerned with how human systems become self-regulating [16]. Processes of self-regulation were seen as essential to a system's ability to maintain its identity in the face of changing environments—certainly a concern for international organizations. In trying to more fully understand the cybernetic process of self-regulation and its relationship to processes of self-organization, Vickers created the concept of an appreciative system. The cornerstone of his appreciative system is what he referred to as acts of appreciation, which hinge, in turn, on what he referred to as reality judgments and value judgments. For Vickers, value judgments, in addition to the everyday sense of goodness or badness, encompass those processes whereby we make something "real" by distinguishing it from a background. Of course, these processes of making distinctions of this kind are tied to the language we use. We then make reality judgments to determine the *state* of that which we have made real by a value judgment. We may thus claim that a room is hot, or, if we have a thermometer, 30 degrees centigrade (a reality judgment), but we must first decide that temperature is something worth marking (and then measuring) in the first place (a value judgment), after which we might then decide that that degree of heat in that situation is good or bad for us. It is the *juxtaposition* of value and reality judgments that Vickers stressed in marking acts of appreciation.

What we need realize in situations in international organizations is that there are diverse processes of making (and then linking) both value and reality judgments. What may be critical features to us (worthy of inclusion in a communication plan) may be invisible to those others who will most be affected by our plan. An understanding of Vickers' acts of appreciation forces us to recognize that our value judgments are tied to the relationships within which we are embedded, and must be understood as such. A reflexive awareness of our own value and reality judgments, and how we make them with others, becomes critical. This is particularly so in international settings, where those "things" (for example, a particular form of a performance appraisal system) we value most may have no currency for those with whom we seek to work.

Defamiliarization

In his work as a philologist and linguistic anthropologist, Becker has become concerned with processes of translation from one culture or language to another [17]. His work with translation points to key issues for communication professionals, who need be aware of what is said, unsaid, and even unsayable in other languages. At the heart of Becker's work, we find his extension of Ortega's articulation of the "deficiencies" (every utterance says less than it wishes to say) and "exuberances" (every utterance conveys more than it plans) of *all* utterances [18]. Becker extends Ortega's work to processes of translation across and within languages (including text and performance). Recognizing our exuberances and deficiencies serves as a guidepost for how we (co)-create what our interactions with others are "doing" or "about." This recognition may involve tacit rules that we, through our observations, claim to structure our "others'" interpersonal ways of being or the implicit meanings we "find" in what our "others" say to each other and to us.

In international organizations, we need realize that much of our activity involves, in one way or another, processes of translation from one language (and its entailments) to another. In order to effect our own translation process in everyday intercultural encounters, whether with a text (or document), video, or another human being, Becker stresses the necessity to suspend or relax the distinctions and assumptions of our own language. This is so whether our language is our "everyday" one, or the professional language with which we work. He urges us to allow ourselves to become "*defamiliarized*," in these intercultural encounters, to allow ourselves to see and hear in a new, and to us, foreign way. A re-familiarization that might ensue would allow us to see *if*, for example, our categories do indeed fit our (shared) experiences with our others, instead of assuming that they do.

Of course, any attempt at suspending assumptions embedded in our own language is going to involve a lot of hard work—particularly since the way that we proceed with our "defamiliarizing" may still involve assumptions of our own language. Such may have been the case for Tom (of the Gothenburg Corporation) should he not have assumed that he knew what Gunnar meant by a "goal." The issue here is not so much to have a correct meaning as it is to initiate a process of learning—a process that is ongoing. Becker, through his processes of defamiliarization and refamiliarization, invites us to think of languaging as a way of living together.

Joint Performance

In writing about the intercultural encounter, Mary Catherine Bateson focuses on assumptions held by many, including communication professionals, that may be at odds with their experience. In particular, she asks us to question the assumption of shared codes, and to look at what might be unfolding in an intercultural encounter as a meeting of different codes [19, 20]. Building on an awareness of the criticality of difference (rather than sameness), Bateson has proposed that we look on our intercultural encounters through the metaphor of "*joint performance*."

Bateson develops this idea through an elaboration of an intercultural experience of her own. During a week-long ritual festival in Iran, she had brought her young daughter to a garden where, as part of this ritual, a sheep was to be slaughtered. Reflecting on the situation, and her "double-identity" within it, as both mother and field-worker, Bateson notes how, as we join in common occasions, we often make up our multiple and interconnected roles as we go along. She stresses the importance of a willingness, indeed of a necessity, to improvise with each other; that is, to participate inventively, in concert with others, in a situation for which there are no previously set rules. Along with this improvisational attitude, there must be, she notes, a continuing and reciprocal openness to clues to the meanings of cultural differences (to each other), and how these newly unfolding differences might play out in performing together. If we take this notion seriously, as I feel that we must, we might find intercultural training programs fostering a way of being with another, rather than like another (recall here Hans' comments about training at Gothenburg).

Indeed, at another level, Mary Catherine Bateson's work points to our "international organizational" experience as not at all bounded by the confines of the organizational setting itself. We might find dinners or festivals to which we are invited or walks in the park as occasions of joint performance that may connect to our organizational worlds differently for each of us.

Dialogue as a Way of Being (Together)

We often speak of the importance of dialogue in any encounter where different perspectives may be displayed. In intercultural situations, this is, of course, the norm. Yet, do we each mean the same thing by "dialogue" when we speak of being in dialogue with another? And, more importantly here, can we create an organizational frame, particularly in an international organization, where dialogic behavior becomes the norm?

Eisenberg and Goodall develop the notion of organizations as dialogues [21]. Their starting point is that dialogue is, above all, balanced communication, and that "to be human means to live in between" [21, p. 40]—balancing an identity (being apart from the world) with a community (being a part of our world). Eisenberg and Goodall recognize three levels of meaning for the term dialogue. It is their third level that concerns us most here.

A first level of dialogue is that of "equitable transaction—where all interactants are afforded "equitable opportunities to speak." A second level is dialogue as "empathic conversation"—with the focus on the participants' ability to see a world from another's perspective. While both of these are obviously desirable, Eisenberg and Goodall note that equitable transaction is more of a precondition, since having a chance to speak says nothing about what the resultant conversation might be like, nor the processes by which multiple perspectives may be reconciled. Further, while empathy has long been considered a hallmark of good interpersonal and intercultural communication [cf. 22], empathic conversation may not take us far enough in our quest for a relational framework, in that, while it offers us a way of orienting ourselves to another, it stops short of telling us how we might be together.

Building on both Buber [23] and Bakhtin [24], Eisenberg and Goodall propose, as a third meaning of dialogue, dialogue as "meeting." Here there is a recognition of dialogue as a framework for a mutual openness to learn and create with each other. They note that within a strictly competitive frame, the idea of empathy (although in an extremely limited sense) can be used strategically and instrumentally (to try to put yourself in the other's shoes to be able to gain a competitive advantage). Dialogues as meeting, on the other hand, are self-validating, in that they are valuable in and of themselves.

While beneficial outcomes may result from this third level of dialogue, those outcomes cannot be tied to what any one of the interactants does. Such outcomes derive from a mutualness, a coupling of desires to learn jointly. Unfortunately, dialogue as "meeting" does not lend itself to the kind of formulaic prescriptions that communication professionals are often called upon to offer (and receive) for how to properly behave in specific intercultural settings. We may talk of a readiness to participate in dialogue, but cannot guarantee that dialogue will result, nor be able to say that such a result could be tied to anything that one person did.

Dialogue as meeting is a cornerstone of a relational framework. As Eisenberg and Goodall note, dialogic process as meeting occurs between individuals, not as something one does *to* another. We must realize that, particularly as communication professionals, it is possible to facilitate conditions from which dialogue might ensue, but we cannot specify beforehand the relationships of those participants (including ourselves) in any dialogue. Indeed, we need to be aware that the way we, as communication professionals, are heard is yet another issue within a relational framework.

Even questioning what we mean by dialogue can, at a process level, embody the very issues this chapter seeks to identify. That is, looking at different meanings of dialogue can illustrate both what those meanings may be (as potentials for joint action) as well as what the process of unconcealing assumed shared definitions may be.

FROM TOOLS TO A RELATIONAL FRAMEWORK

As a system, these four "tools" described above provide a foundation for a relational framework for professional communication in international organizations, in that they all require an awareness of how all that we say, do, and claim to know is embedded in our relationships.

These tools as guides for action in communicative settings in international organizations allow that we 1) need be aware of ours and others' acts of appreciation, including how we and they make judgments of value and reality, 2) need allow ourselves to become defamiliarized, to suspend our usual and comfortable ways of speaking and listening, in order to refamiliarize ourselves to different ways of speaking and listening with different others, 3) not assume that we (ourselves and our others) are operating out of the same code, but pay attention to the joint performances in which we participate, as joint improvisations, and 4) strive to establish mutual readiness for dialogue as "meeting."

Each of these tools requires a form of relational understanding, as each makes clear that all of our organizational knowledge is tied to our relationships within those organizational settings. The Gothenburg example is the beginning of a story of defamiliarization, that may take on a kind of disorientation, waiting to happen in the midst of a joint performance not yet observed as such by the participants. As communication professionals, we need to be cognizant of these tools, as our livelihood is, after all, communication in relationships. International organizations just mark these processes more clearly, just as the need for translation of a text from a different language is more evident than one in your own language.

IMPLICATIONS FOR PEDAGOGY

How we introduce a relational framework is not a separable issue from the content described above. That is, if we feel that a relational framework is needed to appreciate fully issues of professional communication in international organizations, we must extend that relational framework to include how we teach it—whether that teaching is done in a classroom or in a training room. The teacher or trainer, as a communication professional, must also be open to participating in an encounter as guided by processes like the four tools described previously.

This means, above all, that we must be aware of our own acts of appreciation in the learning setting, particularly when those whom we are supposedly teaching may have a different way of understanding those very exercises we use as teaching vehicles. We must be prepared to "defamiliarize" ourselves when we get questions that at first glance seem strange or even misguided, and instead try to understand what experiences may have made the "question" a reasonable one.

Bruce Hyde has noted that the dominant theme guiding much professional communication teaching and practice has been that of persuasion. He offers, instead, as a possibility, a *pedagogy of dialogue,* which has opened up a space (perhaps as dialogue as meeting) for thinking of the classroom (and other learning based settings) in a different way [25]. Of course, there is the irony of perhaps trying to "persuade" others that a pedagogy of dialogue is a good thing, but persuasion is not the intent. It is an invitation to a joint performance, since opening ourselves up to the multiple interpretations possible in any international organization requires a willingness to improvise together, based on the (new) participants' own experiences of and in their own international settings. This of course includes how an invitation to a dialogue, whether in the classroom or the boardroom, is heard.

REFERENCES

1. J. L. Austin, *How To Do Things with Words* (2nd Edition), J. O. Urmson and M. Sbisa (eds.), Harvard University Press, Cambridge, 1962.
2. J. R. Searle, *Speech Acts: An Essay in the Philosophy of Language,* Cambridge University Press, New York, 1969.
3. G. Hofstede, *Culture and Organizations: Software of the Mind,* McGraw Hill, New York, 1991.
4. F. Capra, *The Web of Life,* Doubleday, New York, 1996.
5. M. Elden and M. Levin, Cogenerative Learning: Bringing Participation into Action Research, in *Participatory Action Research,* W. F. Whyte (ed.), Sage, Newbury Park, California, 1991.
6. S. Bråten, The Third Position: Beyond Artificial and Autopoietic Reduction, *Kybernetes, 13,* pp. 157-163, 1984.
7. K. J. Gergen, The Social Constructionist Movement in Modern Psychology, *American Psychologist, 40,* pp. 266-275, 1985.

8. K. J. Gergen, *The Saturated Self*, Basic Books, New York, 1991.
9. E. von Glasersfeld, An Introduction to Radical Constructivism, in *The Invented Reality*, P. Watzlawick (ed.), Norton, New York, 1984.
10. E. von Glasersfeld, Knowing Without Metaphysics, in *Research and Reflexivity*, F. Steier (ed.), Sage, London, 1991.
11. F. Steier, From Universing to Conversing: An Ecological Constructionist Approach to Learning and Multiple Description, in *Constructivism in Education*, L. P. Steffe and J. Gale (eds.), Lawrence Erlbaum, Hillsdale, New Jersey, 1995.
12. J. Shotter, In Dialogue: Social Constructionism and Radical Constructivism, in *Constructivism in Education*, L. P. Steffe and J. Gale (eds.), Lawrence Erlbaum, Hillsdale, New Jersey, 1995.
13. G. Bateson, *Steps to an Ecology of Mind*, Ballantine, New York, 1972.
14. F. Steier, C. Cockrell, and B. Lewis, Rethinking Frames for "Information Sharing" in Organizational Settings: Lessons Learned about a "Lessons Learned Information System" at NASA, in *Platform for Technological Leadership. Proceedings of the 1995 National Conference of the American Society for Engineering Management*, L. D. Richards (ed.), Arlington, Virginia, 1995.
15. F. Steier and E. Eisenberg, From Records to Relationships: Courting Organizational Dialogue at NASA, *Cybernetics and Human Knowing, 4*, pp. 51-58, 1997.
16. G. Vickers, *The Art of Judgment*, Chapman and Hall, London, 1965.
17. A. L. Becker, *Beyond Translation: Essays toward a Modern Philology*, University of Michigan Press, Ann Arbor, 1995.
18. J. Ortega y Gasset, The Difficulty of Reading, *Diogenes, 28*, pp. 1-17, 1959.
19. M. C. Bateson, Joint Performance across Cultures: Improvisation in a Persian Garden, *Text and Performance Quarterly, 13*, pp. 113-121, 1993.
20. M. C. Bateson, *Peripheral Visions*, HarperCollins, New York, 1994.
21. E. Eisenberg and H. L. Goodall, *Organizational Communication: Balancing Creativity and Constraint*, St. Martin's Press, New York, 1993.
22. E. Stewart and M. Bennett, *American Cultural Patterns* (Rev. Edition), Intercultural Press, Yarmouth, Maine, 1991.
23. M. Buber, *Between Man and Man*, Macmillan, New York, 1985.
24. M. Bakhtin, *The Dialogic Imagination*, C. Emerson and C. Holquist (trans.), University of Texas Press, Austin, 1981.
25. R. B. Hyde, Toward a Pedagogy of Dialogue (panel session), Annual Conference of International Communication Association, Chicago, May 1996.

CHAPTER 8

Effective Communication in Multicultural Organizations: A Receiver-Defined Activity

JUDI BROWNELL

Globalization and the increasing diversity of the American workforce offer organizations new challenges as well as new opportunities. Business communication professionals are now in a position to influence and improve business practices by addressing the unique requirements of effective communication in diverse organizational environments [1, pp. 65-79]. Such a responsibility cannot be taken lightly. To create high performing organizations, leaders must understand the impact diversity has on communication, and recognize that fundamental changes in the composition of human resources may necessitate corresponding changes in communication channels, practices, and strategies.

As the workforce becomes more diverse, ensuring communication effectiveness becomes one of management's most critical tasks. Whether communicating to an individual, members of a department, or employees throughout the organization, individuals can no longer take for granted the sources of information on which employees depend, the communication channels they choose, or the meanings they derive from the messages that they receive. Leaders must audit the communication activity within their companies and be ready to respond to individual differences and to changing workplace dynamics. New sets of skills may be required of organizational leaders who seek to promote communication effectiveness as a critical competitive advantage [2, pp. 189-210; 3, pp. 161-168].

This chapter presents a summary of three studies, each of which concludes that the information sources and channels on which non-native speakers depend for organizational information differ significantly from those of native speakers. It becomes apparent that organizational leaders in the coming decades will require a different set of core communication competencies to manage diversity effectively. One of the most critical skills in this new leadership "package" is the ability to identify and respond to individual differences. Effective communication requires 1) increased sensitivity to ones' impact on others, 2) accurate interpretation of meanings, and 3) receiver-based organizational communication strategies. This chapter proposes a theoretical framework developed around the notion of receiver-defined meanings as an appropriate way to view the core competencies of effective business communicators.

Considering the implications of this change in paradigm for business communication educators, I suggest how communication professionals might better prepare future managers to communicate within a highly diverse workforce. The chapter concludes with a discussion of specific communication competencies and teaching strategies that will increase students' awareness of their communication challenges, assess their current competencies, and assist them in developing the skills and strategies required to manage effectively in culturally diverse organizational contexts.

COMMUNICATION AND ORGANIZATIONAL DIVERSITY: THREE CASES

Service organizations, in particular, depend upon effective internal communication as a competitive advantage [4, pp. 337-352; 5, pp. 383-412]. When employee-guest interactions are part of the service itself, guest satisfaction is profoundly affected by the quality and nature of the communication that takes place. These interactions, in turn, are only appropriate to the extent that employees have clearly understood such concepts as quality guest service. Regardless of an employee's cultural orientation or experience, she or he must understand what key organizational concepts mean within the context of the specific job responsibilities. The employee must be able to interact in a coordinated manner with fellow employees in the delivery of quality service. Effective internal communication, at all levels, is central to high performance.

Studies conducted to determine how concepts related to service quality are communicated within hospitality organizations provide insights into cross-cultural differences in the selection and use of communication sources and channels [6, pp. 28-33; 7]. Three full-service, luxury hotels, one in the United States, one in France, and one in England, participated in a study to determine the communication sources and channels employees use to come to their understanding of what quality service means in their particular environment. An instrument was created to audit employee behaviors on this specific dimension and to answer a number of questions related to internal communication practices, among them: 1) *Who* talks with employees about quality service? 2) *How* do employees learn about quality service? and, 3) *What* does quality service mean within the context of the employee's specific position and job responsibilities?

Focus groups, interviews with hospitality employees, and a review of organizational documents related to each hotel's philosophy of service quality provided data that was used to develop items on the questionnaire. A sample is provided in Figure 1. Coded surveys were then distributed to all employees through the human resource department and returned directly to the researcher. Employees were asked to rate, on 7-point Likert scales, the degree to which they relied on various sources of information or used various channels when communicating about quality service concepts. They were provided with an open-ended question and asked to identify the meaning quality service had for them. Demographic information was also requested, including age, years of work experience, years in current position, department, gender, native language, and so forth.

Do differences exist between native and non-native speakers with regard to the most frequently used sources and channels of communication? In each of the three cases, data analysis revealed statistically significant differences between native and non-native speakers on a number of dimensions. It appears safe to conclude that non-native speakers rely on different sources of information and different communication channels than do native speakers. The findings presented in Figure 2 were consistent across all three properties.

1. Who talks with you about quality service?
   - my direct supervisor
   - other managers
   - my fellow employees
   - the general manager
   - guests
   - my family
2. How have you learned about quality service?
   - through pre-arranged conversations
   - through spur-of-the-moment, casual conversations
   - by watching other employees
   - by watching managers
   - by being coached
   - through formal training programs
   - through manuals or other formal written documents
   - through memos or other informal written documents
   - by participating in meetings

Figure 1. Questions asked about service quality.

Sources of information employees depend on for information about service quality

(1) Non-native speakers talk less about service quality with guests than do native speakers.
(2) Native speakers receive more information from conversations with the general manager than do non-native speakers.

How employees learn about service quality

(1) Non-native speakers rely more on written communication than do native speakers.
(2) Non-native speakers rely more on formal, pre-arranged conversations than do native speakers.
(3) Native speakers depend more on meetings for information than do non-native speakers.

Figure 2. Differences between native and non-native speakers in the use of information sources and channels.

In addition, employees were asked to define quality service. When the content of U.S. employees' responses was analyzed, striking differences were revealed in the meanings native and non-native speakers associated with quality service. Eighty-seven percent of the responses clustered around six central themes. A comparison of the percentage of native and non-native speakers who supplied each of the six response types (employees may have provided a response that fell into more than one category) again reveals the need for more closely examining cross-cultural differences in communication. This finding suggests that even when native and non-native speakers are exposed to the same messages, they may interpret the information differently (Figure 3).

Although this study focused exclusively on the communication of service quality concepts, it is likely that differences in the sources and channels native and non-native speakers seek, and the subsequent meanings they derive, exist for other types of organizational knowledge as well. How, for instance, do employees learn what behaviors are rewarded in an organization? How do they develop an understanding of ethical behavior? It is apparent that this finding is important to organizational leaders who strive to create shared meanings within diverse organizational environments, and who strive to affect culture and implement organizational change. Not only is such awareness critical as leaders design organizational communication strategies, greater sensitivity to cultural differences can also assist managers in handling more effectively the daily communication issues that arise within their departments and workgroups [8, pp. 61-82].

Native $N = 102$ Non-native $N = 38$		
Response Category	Percent of Total Native	Percent of Total Non-Native
1. Meet organization's service standards	8.8	68.4
2. Behave in a professional manner	10.8	47.9
3. Anticipate guest needs	33.3	23.7
4. Behave in a friendly, courteous manner	41.2	13.2
5. Provide helpful, attentive service	79.8	18.4

Figure 3. Definitions of quality service.

The recognition that 1) fundamental differences exist in the way native and non-native speakers use formal and informal communication sources and channels, and 2) the meanings these two groups assign to messages may be influenced by cultural orientation, has implications for business communication educators who are interested in preparing future leaders to effectively manage a diverse workforce. It is apparent that employees who are less familiar with the primary language spoken within the organization need special attention in meeting their communication requirements. The remaining sections of this chapter focus on the core competencies required of effective cross-cultural communicators. Subsequent discussion considers what these competencies suggest with regard to the role of business communication educators in preparing students to lead a highly diverse workforce.

IMPLICATIONS FOR BUSINESS COMMUNICATION EDUCATION

To meet the communication requirements of a diverse workforce, organizational leaders must view communication as a receiver-defined activity. That is, communication outcomes are determined by the meanings assigned by the listener or reader, not by the intentions of the speaker or writer. Renewed attention, therefore, must be placed on the receiver's interpretive process as the single most significant factor influencing communication effectiveness.

This shift in focus implies that communication educators must address the new management needs created by fundamental differences in the way employees from other cultures seek out and process information. Those who manage in culturally diverse organizations must be intensely concerned 1) with how listeners construct meanings from the messages they receive, and 2) with identifying strategies that will reach employees with different cultural orientations. Emphasis is shifted from the tasks of the speaker/writer to the interpretive processes of the listener/reader. Consequently, managers must continuously monitor employee interpretations, preferences, and patterns to ensure that messages are received and understood.

Based on this framework, we present a model that provides direction for those concerned with teaching and assessing these critical communication behaviors and that has particular application to understanding cultural differences in organizational settings. We conclude by suggesting teaching strategies that increase students' awareness of the challenges they confront as receivers, assess their current competencies, and assist in developing skills and strategies required to manage in culturally diverse organizations.

Receiver-Centered Communication: Theoretical Foundations

A receiver-centered communication perspective proposes that understanding differences in the perception and subsequent interpretation of messages is at the core of communication effectiveness. Excellent communicators are interested in what goes on in the minds of receivers. They understand that meanings can never be completely shared because no two individuals experience events in exactly the same way. In one respect, organizational communicators are at the mercy of employees who interpret what they hear based on their own assumptions, values, and expectations regardless of what the communicator intended. Employees may believe that a message was sent even when nothing was intended; meanings may be influenced by cultural assumptions and expectations that distort the original communication.

The manager of a large restaurant, for instance, noticed with interest that the formal appreciation dinner he had been providing for his employees didn't encourage the type of social interaction and relaxation he had envisioned. The next year, he decided to try a more informal event and, instead of an evening banquet, he planned an all-day picnic. His hope was that employees and their families would find this a more appropriate way to start off the new year. If he had listened to the talk in the halls, however, he would have realized that employees saw this change as a cost-cutting method. Rather than accepting the explanation that a picnic would be more enjoyable and festive, they believed that the manager was simply trying to get away with spending as little as possible on them.

As we have seen, each employee chooses which stimuli to pay attention to, thereby creating his or her unique reality. In this regard, communication is a creative process, with each person an active participant in determining the outcome. Employees not only select different stimuli, they also control the amount of information they process. These characteristics suggest that potentially important verbal and nonverbal cues are often missed. On occasion, meanings may even be elicited in situations where messages were never intended.

When individuals have similar backgrounds and share a common body of experience, communication can be facilitated more easily than when cultural diversity brings individuals with very different perceptual filters and values together. The concept of high and low context is relevant to organizational communicators [9]. In high context cultures, communicators share an experiential base that can be used to assign meanings to messages. If an employee passes a colleague in the hall and says, "Hey, the dragon is breathing fire today!" he expects to be understood because of a shared background. Both parties understand that the sales and marketing manager, Pete Bell, is affectionately called "the dragon" by members of his department. They know that sales were down in the eastern region, and that Pete just held a meeting to discuss second quarter results.

Low context cultures, on the other hand, provide little information on which to base common understandings and so communicators must be explicit. In the situation above, a communicator who recognizes the need for providing the entire context within the communication itself might say, "Hey, Pete Bell is really upset about the drop in sales in the eastern region. Better be careful!"

It is most productive to view diverse organizational environments as low context cultures, since it is unlikely that employees share a common background apart from the organization itself and, as a result, the meanings they assign to key organizational concepts may be unique. For instance, an Arab's understanding of quality service may differ significantly from the German, Japanese, or Puerto Rican point of view. In one culture, a formal and detached approach will little eye contact may be preferred; in others, service personnel would be viewed as inattentive and rude unless they behaved in a friendly and outgoing manner. These differences in expectations and interpretation are likely to have a significant impact on how service is delivered and subsequent guest perceptions of service quality. In such instances, ideas must be expressed fully and explicitly to ensure similar meanings are elicited by individuals with different cultural orientations.

When organizational leaders are charged with creating shared meanings, then the notion of a third culture becomes useful [10]. The third culture concept proposes that communicators are most effective when they work to create clear and consistent messages that are unique, thereby developing a separate and distinct organizational culture. Assuming that individuals from China and the Bahamas may bring entirely different perspectives on quality service to their jobs, the most effective organizational communicators are those who focus on creating meanings that are organization-specific. The third cultural approach, then, explains how employees with very different backgrounds can be encouraged to develop shared meanings within the context of the work environment.

If we bring together the two lines of thought developed so far, we conclude that employees from different cultural backgrounds will select and filter information in different ways. Their channel choices and primary sources of information may be dissimilar, potentially leading to misunderstandings and inadequate information. Therefore, the most effective leaders work to understand cultural differences and to accommodate them by paying attention to the channels and sources employees choose. It is dangerous to assume that any given communication will reach all employees and, subsequently, have the desired impact. One method of gathering information about daily practices is through a communication audit, a survey designed to provide complete and concrete information on employees' perceptions and communication behaviors.

In addition, the meanings employees assign to key organizational concepts are affected by their cultural backgrounds. To ensure the highest probability that meanings will be shared, communicators are advised to assume low context environments; that is, information must be explicit and complete. The third culture concept provides a framework for creating organizational cultures that exist separate from the assumptions and expectations of any one particular cultural orientation or viewpoint.

The following section explores the specific receiver-defined communication competencies leaders might bring to their organizations that will prepare them to communicate effectively in diverse environments.

Communication Competencies for Diverse Organizational Environments

The most effective organizational communicators are those who attend to individual differences on the one hand, and who work to develop shared organizational meanings, separate from culture-specific understandings and assumptions, on the other. In both efforts, receiver-based skills are critical. This section reviews the three competencies that distinguish effective communication in diverse organizational environments. Specifically, skills relate to 1) self-monitoring, 2) empathy, and 3) strategic decision making. After briefly describing each skill, I suggest instructional methods that business educators can use in fostering the development of each. These sample classroom activities are provided at the end of the chapter.

Self-Monitoring: Theory and Teaching Implications

Self-Monitoring: Theory—Effective receiver-defined communication begins with self-monitoring. Self-monitoring refers to a communicator's awareness of how his or her behavior affects another person, and his or her willingness to modify this behavior based on knowledge of its impact. Snyder defines self-monitoring as "self observation and self control guided by situational cues to social appropriateness" [11, p. 526].

High self-monitors are particularly sensitive to individual differences and the needs of others. They attend to subtle changes in communicative behaviors such as vocal quality, mannerisms, eye behavior, and body posture. High self-monitors gather information about an individual's feelings, attitudes, and level of understanding and address them accordingly. Although there may be cultural differences in the meanings such behaviors have for communicators, high self-monitors are able to detect key responses and identify potentially troublesome communication situations. They then respond appropriately, modifying their own behavior in light of their partner's reactions.

A high self-monitor, for instance, might set an appointment with her supervisor to discuss ideas for an important proposal. As the discussion unfolds, the supervisor may glance repeatedly at her watch or papers on her desk. Her nonverbal communication may not always be consistent with her statements, and she may fidget or appear distracted. The high self-monitor would 1) observe this behavior as a significant part of the communication situation, and 2) adapt to what she perceives in some way—either by making her conversation shorter than planned, or even asking her supervisor whether it would be more convenient to meet at another time. In contrast, a low self-monitor would move forward with the original plan, focusing on her own ideas rather than on the reactions of her communication partner.

The tendency for high self-monitors to adapt their behavior to the needs of the situation, as determined by their partner's response, make them particularly effective in diverse organizational environments. These individuals are able to identify employees who are confused and who are not responding appropriately to organizational messages, and modify their behavior to fit the requirements of the specific setting and individual. They also readily adapt to third culture communication requirements.

Low self-monitors, on other hand, rely heavily on their cultural background, values, and individual style in responding to communication events. Their behavior is relatively inflexible and consistent from one situation and one individual to the next. Because of this tendency, low self-monitors have difficulty recognizing and accommodating cultural differences. They may find it stressful to collaborate with colleagues who don't share their assumptions and beliefs, and they are more likely to resist adapting their perspective to organizational demands.

(2) Self-Monitoring: Teaching Implications—The business communication classroom provides a safe environment in which students can discuss their own communication behavior and the impact it has on others. Self-awareness activities encourage students to share their personal experiences, explore communication options, and discuss the degree to which they self-monitor in the various situations they regularly encounter. Values clarification activities are particularly appropriate for facilitating productive, non-threatening discussions. Such simple questions as, "Are you more a living room or a screened-in porch?," "Are you more of a rose or a daisy?," or "Are you more here or there?" serve as a catalyst for discussions regarding students' self-perceptions on dimensions such as openness,

formality, conformity, or sociability. Once again, different interpretations may emerge as students share their unique viewpoints.

As they work to become more skilled self-monitors, students direct their focus outward to identify how their behavior is affecting their partner. They solicit feedback regarding how their actions are perceived. Exercises that depend on a relatively high level of disclosure require a supportive and open classroom environment. Students must understand the principles of constructive feedback and appreciate individual differences for such exercises to be positive and productive.

Empathy: Theory and Teaching Implications

(1) Empathy: Theory—Empathy is a key ingredient for understanding and accurately interpreting the meanings conveyed by those whose cultural orientations differ [12, pp. 185-200; 13, pp. 10-15]. Empathy enables the receiver to go beyond the literal meaning of a message and consider the communicator's feelings, values, assumptions, and needs. Two relevant dimensions of empathy for cross-cultural communication are the cognitive aspect and the perceptive aspect.

Cognitive empathy involves taking the role of the other person and making an effort to view a situation as she sees it. If, for instance, an employee has nothing to say when asked for reasons why she should be given a raise, you might think, "She takes no initiative and therefore doesn't deserve the raise." In looking at the situation from the employee's perspective, however, you might reconsider. In this case, you would reflect upon the fact that she is Asian and may be feeling uncomfortable and alienated. You would recognize how culture plays a role in determining her response and in shaping her interpretation of the situation. Given these insights, an effective communicator may reframe the situation to elicit the desired response in a more appropriate and non-threatening manner. A receiver-centered communicator might ask what she enjoys most about the job, what goals she has set for herself, or how others have helped her to succeed.

Empathy also involves sensitivity to nonverbal communication and the ability to interpret connotative aspects of language. This skill, called the perceptive aspect of empathy, requires taking into account nonverbal cues as well as other individual and situational factors. When a front desk employee makes no eye contact with guests, has a slumped posture, and responds to guests in a monotone, the empathic manager immediately recognizes a problem. The fact that such nonverbal indicators may vary with cultural orientation makes perceptive empathy even more challenging in diverse environments.

Although the skills of empathy can be learned, using those skills effectively depends as much on attitude as on behavior. The true empathic communicator likes people and has a sincere desire to understand them better. Empathy requires a responsiveness to both anticipated and unexpected messages from others. It is also important to recognize that empathy requires reciprocity; the other person must be willing to reveal his or her emotions. If individuals manipulate all aspects of their communication, empathy becomes impossible.

You may have encountered a situation where a friend has experienced a tragic event or a disappointment. If your friend refuses to reveal his feelings to you—if he tells you that he is just fine when in fact he is upset or fearful, and behaves in a manner that supports his statements—it will be nearly impossible for you to demonstrate empathy. Such examples occur in the workplace when employees are reluctant to acknowledge their disappointment at not getting a promotion, their anger at being asked to work overtime, or their frustration when their supervisor does not fully acknowledge their contribution to the success of an important project.

Keep in mind, too, that empathy is constructed by the receiver, who can never completely put herself in another person's position. Empathy, at best, is always one person's good guess as to what another is experiencing.

(2) Empathy: Teaching Implications—The cognitive and perceptive aspects of empathy are two quite different skills. The cognitive component of empathy requires open-mindedness and the development of a sincere respect for individual differences. Students must work to withhold judgment, to understand before they evaluate, and to value differences.

The cognitive aspect of empathy is best developed through exposure to a wide variety of cultural perspectives and frank discussion of differences. Students come to recognize that there are multiple perspectives from which any situation can be viewed, and they learn to identify their own agendas and bias as well as to recognize the beliefs and expectations that underlie the viewpoints of those who represent other cultures. As they become more objective, students find it easier to listen to others whose perspectives and opinions differ from their own. They are less concerned with presenting their own viewpoint and more interested in accurately understanding their partner's position.

Behavioral empathy can be developed as students become more aware of nonverbal behavior as a critical part of any communication. Accurately interpreting nonverbal messages requires recognition of cultural differences as well as sensitivity to the cues themselves. Differences in interpretation of nonverbal behavior across cultures is striking, and students become more sensitive communicators by taking into account how culture influences eye behavior, body posture, facial expression, and movement.

Such sensitivity can be developed not only through class and small group discussion, but also through role plays, observation, and analysis of intercultural interactions. The more students become aware of differences in real-world settings, and bring these examples into the classroom, the more aware they will become of how nonverbal behavior affects their own communication encounters.

Strategic Decision Making: Theory and Teaching Implications

(1) Strategic Decision Making: Theory—Effective organizational communicators, as we have seen, must use what they know about effective cross-cultural communication and workforce diversity to plan more comprehensive communication strategies. At the macro level, empathy translates into an ability and willingness to take into account cultural differences in the design of organizational communication systems. Self-monitoring becomes the monitoring of organizational communication efforts in light of outcomes to ensure that key messages are eliciting intended meanings for all employees, regardless of cultural orientation. Leaders must ask themselves, "How am I being understood by employees at all levels of the organization? How can I adapt communication sources and channels so that my meanings are accurately interpreted?"

It is not enough for organizational leaders to be eloquent speakers and effective writers. Whenever important messages need to be communicated throughout the organization, receiver-centered approaches are essential to making wise strategic communication choices. Leaders who are flexible and who adapt to the unique needs of their workforce by considering employees' perceptual filters, beliefs, and assumptions in designing organizational communication strategies will have a clear competitive advantage.

The housekeeping manager of a four-star hotel, for instance, may be bright and articulate, but if her Hispanic staff has difficulty with the English language, she may need to take extra steps to meet their specific communication requirements. Written directions accompanied by pictures and frequent one-on-one communication may be more appropriate than a high-power traditional training seminar. A creative sales staff, however, may benefit more from an interactive setting where they have opportunities to share ideas and learn from one another. The most effective organizational leaders help their managers to respond to the unique needs of each department while pursuing a common goal.

Strategic decision making in tomorrow's high-performing organizations implies that the communication sources and channels used to reach organizational members, as well as the substance of the messages conveyed, are mindfully selected. It is clear that cultural orientation affects the ways in which individuals seek, use, and process information. Leaders can no longer be satisfied with current communication structures, but must continuously monitor and improve the delivery systems through which employees learn about key organizational values and concepts [14, pp. 8-12; 15, pp. 131-150; 16, pp. 140-161].

Receiver skills can be taken to a macro level through such tactics as communication audits or inventories like the one described earlier in this chapter. Organizational leaders who are interested in improving communication effectiveness begin by assessing their audience's backgrounds, preference, beliefs, and assumptions. Comprehensive assessments, often in the form of a survey, provide essential information about employee preferences and practices that can then be used as a basis for strategic communication decisions. Only when managers understand how culture affects employees' selection and interpretation of organizational messages are they in a position to design and deliver effective communications.

(2) Strategic Decision Making: Teaching Implications—Students accustomed to thinking in terms of face-to-face communication must recognize the challenges leaders confront as they plan organizational communication strategies to reach all members of a diverse workforce. Cases designed to highlight issues in workforce diversity in mid- to large-sized organizations facilitate this goal. Students can be asked to examine alternative strategies and to create communication systems within the context of a specific case. *Resistance to Technology,* presented below, is one such example.

Case: Resistance to Technology

You are the General Manager of a hotel that has recently made a commitment to take advantage of technological advances by acquiring a new computer system that will, among other things, link all departments. Unfortunately, most of your employees see this new system as a threat, and several of your managers are having difficulty introducing the equipment. In fact, you suspect that even at the management level there is not one hundred percent support for this decision—even though they were involved in the discussions that led to the purchase.

For example, you discovered that your reservations manager, who has been with your company for over twenty years, just received a shipload of computer equipment for the front desk but immediately had the boxes placed, unopened, in the housekeeping storage closet. When you inquired about his plans to introduce the new system, he told you that the front desk staff was doing just fine and that he'd present it "when the opportunity arose." He is European, as are a number of the front desk staff, and values the personalized tradition he has worked hard to establish. You know he fears that the technology will replace the "human" aspect of service. He also mentioned that many of the Asian employees find the computers so fascinating, and are already so skilled, he fears they will devote too much of their attention to the hardware. Your housekeeping manager was even more resistant, explaining that her staff is Hispanic and most of them have difficulty with English anyway. She said that asking them to log in on a computer was the "most ridiculous thing" she'd heard since she began working at the hotel.

You suddenly realize that the challenge is greater than you imagined. Attitudes range from passive acceptance to hostility. No one is sure how the technology will influence his or her job, and training isn't scheduled to begin for three weeks. There is confusion, anxiety, and a general lack of support.

Case questions become important as they direct students' thinking and provide a forum for the exchange of ideas. Questions may vary according to the sophistication of the group and the objectives of the assignment. Given the situation presented above and the goal of encouraging students to think more strategically, the following questions might be provided.

1. Fully explain the organizational problem presented in this case.
2. As the General Manager, design a communication strategy to address this dilemma.
   a) What meanings do you need to convey?
   b) How would you use the various sources and channels to communicate with your employees?
   c) How would you sequence these messages?
3. How would you account for workforce diversity in the sources and channels you recommend?
4. How would you determine if your strategies were successful?

The most useful cases require students to *make choices* regarding the most appropriate sources and channels of information and to adapt their strategies to the needs of various groups. A discussion of communication audits and other techniques for assessing the effectiveness of communication systems may stimulate students to consider the challenges involved in creating shared meanings among employees from different cultural orientations. Students may also be willing to share their personal experiences as members of an organization and the kinds of messages they received about key concepts and values.

CONCLUSION

Findings from the three studies summarized earlier suggest that organizational leaders must carefully assess communication sources and channels before designing the communication strategies through which key organizational messages are conveyed. When communication is viewed as a receiver-defined activity, organizational leaders become more aware of the need for focusing on employee differences at both the individual and organizational levels. Through self-monitoring, empathic listening, and strategic communication planning and decision-making, leaders increase the likelihood that organizational messages will be received and understood.

Whether designing strategies to complement the communication of organization-wide messages, or training a front desk employee, cultural orientation is a key communication variable. Business communication educators, as they prepare individuals for leadership roles, may well find that the skills of self-monitoring, empathy, and strategic decision making are high leverage activities. Emphasizing the challenges and complexities of effective internal communication in diverse environments doesn't make it more difficult; it may, however, motivate organizational leaders to make communication decisions in mindful ways. Equipped with a receiver orientation toward their communication tasks, organizational leaders will undoubtedly discover that their attention to cultural orientation, and the accompanying adjustments such a perspective requires, will enable them to lead a diverse workforce into the next century with renewed commitment and focus.

Teaching Strategies for Developing Receiver-Centered Communication

Self-Monitoring

1. Show a movie or television clip that provides students with an example of self-monitoring. Students may provide additional examples.
2. Ask students to form small groups of approximately four to six members each. Request that each person take a turn discussing a situation where self-monitoring was required and assessing the outcome.
3. Check to ensure that all class members understand the difference between high and low self-monitors. On a continuum from very high to very low represented by a scale from 1 to 10, ask students to place themselves on the scale. Request volunteers to share their self-perceptions and briefly describe why they placed themselves at a particular point.
4. Discuss how a particular role relationship or situation affects self-monitoring.
5. Discuss how self-monitoring differs between oral and written communication. In what ways do business writers self-monitor?
6. Identify examples of an individual's self-monitoring as described in literature—plays, short stories, or novels. How did the self-monitoring change the dynamics of the communication encounter?
7. Provide students with short role play situations. Stop the action at a critical point, and request that one of the participants describe the impact his or her behavior is having on the other person, and how he or she is responding to that knowledge. A performance appraisal dialogue where an employee is receiving constructive criticism or a stress employment interview situation are both effective in illustrating the importance of self-monitoring.

Empathy

1. Choose an emotionally charged topic, such as drug testing, no smoking requirements, or other personnel policies, and ask members of the class to describe their positions and reactions. When differences occur, discuss the past experiences and expectations that led to each position. Determine if cultural orientation was a factor in any of these instances.
2. Identify a famous person who demonstrates cognitive empathy. Discuss how this empathy was conveyed and its impact on the other person.
3. Ask members of the class to identify a conflict situation that might arise from lack of cognitive empathy, such as a soft-spoken woman attempting to voice her opinion in a heated board room discussion. Request volunteers to role play

the conflict, first when neither party shows empathy and then again when one person demonstrates cognitive empathy. Discuss the outcomes.

4. Find examples of how business writers demonstrate cognitive empathy, not only in selecting content but also in manipulating style and tone.
5. Discuss some common nonverbal cues and ask members of the class to describe the particular meanings each conveys. Which ones vary according to culture? Eye contact, for instance, may be interpreted differently in different cultural contexts. What challenges might these differences present for organizational communicators?
6. Discuss the ways in which behavioral empathy is important to managers who want to empower their employees. Ask students to name other contexts and work situations where behavioral empathy becomes particularly important.
7. Ask class members to stand for thirty seconds facing a partner and studying his or her appearance. Then, ask participants to turn their backs to one another and quickly make five "changes" in appearance—unbuttoning a shirt, taking off a piece of jewelry, and so forth. When participants face each other again, their goal is to see how quickly each person can then identify all changes that his or her partner made. Ask students what they learned about their perceptiveness regarding appearance.

Organizational Strategy

1. Ask students to share first-person experiences regarding the effectiveness of organizational communication, describing both the message and the channels that were used to inform employees. Compare the strategies used by various organizations and the effectiveness of each strategy.
2. Ask students to identify several messages that are particularly difficult to communicate to a diverse workforce. Discuss the options available to organizational leaders in communicating these messages. Examples of such messages might include changes in work hours, layoffs, or other new policies.
3. Communication audits are carefully structured instruments designed to gather information about how employees perceive communication in their organization. Identify a specific organization and a specific communication problem, and ask students to create a survey that will provide the information needed to create a more effective communication system. An example might be low morale at an auto factory experiencing layoffs, or the perception in an insurance company that women and minorities are discriminated against.
4. Cases provide effective catalysts for examining issues in organizational communication strategy. Provide class members with a case that requires communication planning and problem solving. Ask students to work on the case in small groups, and then request that each group share its plan with the rest of the class. Compare the plans from various groups and ask each group to justify its

decisions. Cases can be brief, as is the one provided earlier in this chapter, "Resistance to Technology."

REFERENCES

1. D. A. Waldman, A Theoretical Consideration of Leadership and Total Quality Management, *Leadership Quarterly, 4*:1, pp. 65-79, 1993.
2. J. A. Hess, Assimilating Newcomers into an Organization: A Cultural Perspective, *Journal of Applied Communication Research,* pp. 189-210, May 1993.
3. B. J. White, Developing Leaders for the High-Performance Workplace, *Human Resource Management, 33*:1, pp. 161-168, Spring 1994.
4. B. P. Niehoff, C. A. Enz, and R. A. Grover, The Impact of Top-Management Actions on Employee Attitudes and Perceptions, *Group & Organizational Studies, 15*:3, pp. 337-352, 1993.
5. B. Schneider, The Climate for Service: An Application of the Climate Construct, in *Organizational Climate and Culture,* B. Schneider (ed.), Jossey-Bass, San Francisco, California, pp. 383-412, 1990.
6. J. Brownell and D. Jameson, Getting Quality Out On the Street: A Case of Show and Tell, *Cornell Hotel & Restaurant Administration Quarterly, 37*:1, pp. 28-33, 1996.
7. J. Brownell, Quality Across Cultures: A Study of Communication Sources and Channels, *Working Paper,* School of Hotel Administration, Cornell University, Ithaca, New York, 1996.
8. H. Greenbaum, E. Holden, and L. Spataro, Organizational Structure and Communication Process: A Study of Change, *Group and Organization Studies, 8*:1, pp. 61-82, 1983.
9. D. Victor, *International Business Communication,* HarperCollins, New York, 1992.
10. T. D. Thomlison, Intercultural Listening, in *Listening in Everyday Life: A Personal and Professional Approach,* D. Borisoff and M. Purdy (eds.), University Press of America, New York, 1991.
11. M. Snyder, Self Monitoring of Expressive Behavior, *Journal of Personality and Social Psychology, 30,* pp. 526-637, 1974.
12. T. Bruneau, Empathy and Listening, in *Perspectives on Listening,* A. D. Wolvin and C. G. Coakley (eds.), Ablex, Norwood, New Jersey, pp. 185-200, 1993.
13. R. Blesius, The Concept of Empathy, *Psychology: A Journal of Human Behavior, 26*:4, pp. 10-15, 1989.
14. R. M. Donnelly, The Interrelationship of Planning with Corporate Culture in the Creation of Shared Values, *Managerial Planning, 32*:6, pp. 8-12, 1984.
15. E. M. Eisenberg and P. Riley, Organizational Symbols and Sense-Making, in *Handbook of Organizational Communication,* G. M. Goldhaber and G. A. Barnett (eds.), Ablex, Norwood, New Jersey, pp. 131-150, 1988.
16. S. A. Sackman, Culture and Subcultures: An Analysis of Organizational Knowledge, *Administrative Science Quarterly, 37,* pp. 140-161, 1992.

PART THREE:
Teaching and Research in International Professional Communication

CHAPTER 9

Cultural Biases in Intercultural Business Communication Courses and How to Avoid Them

S. PAUL VERLUYTEN

In the last few years the demand for training in intercultural communication has been growing at a vastly accelerated pace throughout the world. In step with this, the number of scholarly publications (books and articles) and the number of textbooks for students is also increasing rapidly. Almost every year we witness the publication of fine introductory textbooks that intercultural trainers can use in their classes, such as David Victor's *International Business Communication* [1], Ferraro's *The Cultural Dimension of International Business* [2], or Varner and Beamer's *Intercultural Communication in the Global Workplace* [3]. There is also a healthy and stimulating debate as to the appropriate *teaching methods and materials* for courses in intercultural communication in business.

Yet when I use these and other materials for my own classes and seminars in intercultural communication, I am often left with the distinct feeling that they are not optimally suited to the audience I am using them with. The reason for this lies in a discrepancy between the cultural background of the authors of the textbooks and the trainer on the one hand, and the cultural background of the trainees on the other.

The vast majority of the published material is American, i.e., written by Americans and/or published in the United States. A much smaller proportion of the publications is (Northern) European; publications from other parts of the world are very rare indeed. The trainers themselves, also, are still mainly Westerners (Americans or Europeans), even when the training is carried out in Asia or other parts of the world.

In the past few years I have personally taught full-length intercultural communication courses in a university setting to Belgian, American (U.S.), Filipino, and Thai undergraduates. Not surprisingly, my feeling was that the teaching materials I used (such as the books quoted above or Adler's *International Dimensions of Organizational Behavior* [4]) were well adapted to my American audiences, less optimally adapted to Belgian and Filipino students, and even more poorly adapted to my Thai students.

In the remainder of this chapter, I shall assume the following hypothesis to be true: the more the cultural background of the trainer and the origin of the teaching aids and methods he/she uses differ from the cultural background of the trainees, the higher the likelihood that the intercultural communication class will not be optimally adapted to the trainees.

CULTURAL BIAS

The adaptation described above finds its origin in the fact that both the trainer and the trainees are *biased* by their respective culture of origin in their approach to the training program and their expectations from it.

Obvious though this may seem, in reality the compatibility of the trainers' and the trainees' cultural backgrounds and its impact on the success of a training package in intercultural communication have received little attention up to now. Most authors of training materials and textbooks ignore the issue completely. They discuss culturally determined differences in fields such as management styles, marketing strategies, business communication practices and the like, and they emphasize the obstacles this may create in the global economy; yet they seem to be blind to the idea that they themselves may exhibit such culturally determined features, and that these may equally well be an obstacle to providing adequate intercultural training on a global scale.

Perhaps the assumption is that trainers in intercultural communication are themselves free from cultural bias precisely because they are specialists in cross-cultural differences. Of course (and fortunately so), extreme forms of cultural prejudice and ethnocentrism are likely to be absent from the publications and practices of intercultural specialists. I reserve the term *cultural bias* for milder and more hidden, often unconscious forms of cultural prejudice. These often arise from the assumption that something *seems so obvious and natural that it must be universal.*

It is possible to find examples of cultural bias in the finest textbooks and other publications. I examine two examples below, Ferraro [2] and Varner and Beamer [3]. I selected these two books precisely because, in my view, they are among the *best* in avoiding cultural bias to a large extent. If I can show that even these are not immune to it, then we may assume that nearly all existing textbooks and other teaching materials are subject to (Western and/or American) cultural bias.

Throughout his book, Ferraro shows deep respect and warm appreciation for cultural features from other parts of the world that are quite foreign to Western culture, such as polygamy [2, pp. 35-36] and others. Yet in several passages, cultural bias appears. For instance, Ferraro uses the term "traditional peoples" to refer to certain non-Western cultures [2, p. 92], implicitly suggesting an idea of evolution in which certain peoples lag behind (are more backward?) as opposed to others. More subtle, and therefore perhaps even more interesting, is the following passage:

> Middle-class North Americans have worked out one set of cultural patterns, whereas Indonesians may have developed a radically different solution. In most cases, one solution is probably not inherently more rational than another. They simply represent different answers to similar societal problems. Only after we understand why we do the things we do can we appreciate the internal logic of why other, culturally different people do what they do [2, p. 152].

In this passage Ferraro pleads in favor of cultural relativism, the opposite, in fact, of cultural bias. Yet even here I detect at least two instances of (Western) cultural bias:

1. The standard by which cultural features are judged and found equal in worth by Ferraro is *rationality* (line 3). Rationality is highly rated in Western value systems, but hardly universally so. In fact some solution the Indonesians have developed may be less rational than the North American way of doing things, but this might be because being rational is not an important yardstick by which the solution is measured in that part of the world. Esthetic, religious, or other considerations may have higher priority than being rational.
2. Similarly, every respectable culture is assumed by Ferraro to have an *internal logic,* which seems to refer to the internal consistency of the different features of a culture. Here again, it is not certain that logical consistency is of high value in all cultures. Many South East Asian cultures adopt features from different, mutually contradictory ethical and religious systems in ways that are most certainly *not* logically consistent.

The only textbook authors I am aware of who explicitly address the issue of their own possible cultural bias are Varner and Beamer:

> We have worked very hard to avoid specific cultural viewpoints in this book. The framework we develop applies to all readers regardless of their native cultures. This book is for anyone from anywhere around the globe who wants to improve intercultural business communication skills [3, p. vii].

In other words, Varner and Beamer claim to have produced a textbook that is free from cultural bias. Here again, my purpose is not to blame the authors for not quite living up to this claim, but to show how difficult it is, even for the most open-minded people, to rid themselves of cultural bias. It is undeniable that Varner and Beamer put much care and effort in being open-minded and unbiased. For instance, in their discussion of the status of women in different cultures, they write that, when judging inequity between men and women to be morally wrong, "'wrong' is a culturally based attitude" [3, pp. 96-98]. Not many American (or Western) women would be willing to share such a high degree of cultural relativism!

Yet cases of cultural bias subsist throughout their book, as the following examples will show.

1. One of the five general areas that are identified for asking questions about cultures is: "what do [people in this culture] consider achievement" [3, p. 5]. But the concept of *achievement* is so specific to American culture (and possibly some other variants of Western culture) that

   > [. . .] French and many other modern languages have no adequate equivalent for the English "achievement" [5, p. 35].

   What I am claiming therefore is not, as the authors would suggest, that different cultures have different ways of defining what *achievement* means for them; I claim that the concept of achievement itself is likely to be irrelevant in describing and understanding many cultures in the world.
2. In their discussion of corporal punishments in some Muslim countries, Varner and Beamer write:

   > [i]ts punishments sound medieval: amputation of hands of thieves, stoning of unvirtuous women. Not surprisingly, they are effective [3, p. 83].

   I wonder what a Muslim must make of a one-page passage intended to describe Islam, where sharia-inspired punishments such as those mentioned above are singled out (in fact, very few Muslim countries have adopted these) and said to "sound medieval," while in the discussion on Christianity or throughout the various discussions on Western culture in the book, no mention is made of the fact that the most powerful Western nation electrocutes, poisons, asphyxiates with toxic gases, or puts before a firing squad some of its citizens convicted of crimes. Would Varner and Beamer be willing to apply the same qualification of "medieval" to these U.S. practices? (Note that there may also be a case of *historical* bias, with the use of the word *medieval.* Our biased perception is often that the Middle Ages are an age of ignorance, cruelty, and barbarism, as opposed to Antiquity and the Renaissance and classical periods. Historically, this view is not tenable any more today (see, for instance [6]). Many forms of harsh corporal punishment in European countries are more typical of the Renaissance and classical periods than of the Middle Ages).

Curiously, also, the medieval sounding punishments are ascribed to two enemies of the West (Iran and Libya), but Western ally Saudi-Arabia is not mentioned.

Finally, as a European reader opposed to corporal punishments and the death penalty in my own culture, I am surprised by the claim that such punishments "are effective." To my knowledge, there is no hard evidence to suggest that Muslim countries where the sharia is in effect have a lower crime rate, or less adultery, than some neighboring Muslim nations where such sharia-type punishments are not applied. What I detect behind this way of presenting things is the American belief that harsh *individual* punishment (being "tough on crime") is the preferred way to deal with crime. This runs contrary to a prevalent European analysis of the problem of crime, where more emphasis is placed on the *social* factors that cause criminal behavior.

In brief, I am convinced that the text as it appears in Varner and Beamer's book could only be written by Americans, and not, for instance, by a European or a Muslim: it exhibits some typical instances of American bias.

3. Other, sometimes minor instances of cultural bias could be mentioned. I believe it is culture-specific to assume that "in all cultures, managers have reached their positions through hard work" [3, p. 136]; I personally know managers in various cultures who have *not* reached their position through hard work, and I know cultures where "hard work" is not highly valued.
4. The very terminology used will sometimes prevent Varner and Beamer's book from being understood by everyone, everywhere. When they write that some prime ministers' education "ended at level 12" [3, p. 107], many readers throughout the world, including myself, don't have any idea what that expression says about the age at which the prime ministers left school.
5. The authors have a habit of referring to a manager or another professional with the pronoun "she." The question here is not whether this is good or bad practice. What is certain is that it reflects culture-specific priorities (American or Western) and a usage of the English language that are not universally shared.

With the discussion above, I wanted to show that intercultural specialists, while they may be less prone to cultural bias than most, are not immune to it. Any trainer in intercultural communication must therefore be constantly on the look-out for possible cases of cultural bias in his/her own behavior and in his/her training materials.

I am not a priori convinced that cultural bias can be eradicated completely, that writing a culture-neutral text is possible. In fact, I tend to think the opposite is true. Examining this interesting question in detail is beyond the scope of the present chapter. I do believe, however, that anyone should aim at eliminating instances of cultural bias *as much as possible.* Becoming aware of its many guises is the first step toward this.

The remainder of this chapter consists of a (non-limitative) list of factors which relate to the teacher-trainee relation in intercultural communication classes and where cultural bias may frequently appear.

CULTURAL FACTORS THAT MAY INFLUENCE THE TRAINER-TRAINEE RELATION IN INTERCULTURAL COMMUNICATION CLASSES

The Need for Intercultural Awareness and Training as Felt by the Students

Several American textbooks start with an attempt to convince the reader that intercultural awareness is important in the contemporary business world [3, 4]. It is a well-known fact that some Americans are not, as yet, convinced that investing in intercultural skills is worth the money. For decades, many American companies have produced for their domestic market only, and the awareness that there is an increased need to go global and, therefore, to study intercultural issues, is not always present.

While such a passage is functional in a textbook intended for American readers, I doubt its usefulness when addressing European students. Many European countries, and in particular the smaller ones, have earned close to (or over) 50 percent of their GNP through exports for many years, and doing business across the national borders is neither new, nor is it necessary to convince people that certain skills are needed to do business internationally.

On the other hand, a common attitude among Europeans in some countries remains that such skills do not need to be taught in a formal training program. Many would feel that intercultural skills are a question of individual talent combined with exposure "in the field." In a survey that was carried out by one of my students (Els Grieten) in 1993 among Belgian (Flemish) companies that do business with Poland, Hungary, or (the former) Czechoslovakia, not one of the companies appeared to provide formal intercultural training to its staff. Most would answer that it is through experience that one learns how to deal with partners from foreign countries. The situation is probably different from one country to the next. While the demand for intercultural training programs is still relatively small in Belgium, for instance (even if growing rapidly), it is already much higher in neighboring Holland.

For a European textbook, or for a trainer who is teaching a European audience, it is therefore probably advisable to start with an introduction where he/she shows that *formal training* in intercultural communication can be an invaluable asset that may prevent the individual from making serious mistakes, and which may save the company a lot of money (this is done, for instance, in [7]). With American students, on the other hand, the first task may be to convince them that intercultural skills in themselves are necessary, as in [4] or [3]. Once this is achieved, there may be little need to convince American audiences that formal training can be helpful in this field.

What about non-Western audiences? For several reasons, I believe that with students from some Asian cultures,[1] there is little need to emphasize either the necessity of intercultural awareness and skills in the global economy, or the need for formal training in that field. In fact, Asian students will be readily convinced of both at the outset. Those Asian countries which have been economically successful in recent years owe this success, to a large extent, to the conquest of foreign markets. In Japan or Thailand, it will be difficult to find anyone in business who is not convinced that skills in treating with foreign partners are important. Moreover, the eagerness to gather knowledge combined with the respect in which teachers are held in many Asian countries is such that there won't be much need to convince the students that learning about intercultural skills in a class setting is helpful.

As the previous discussion suggests, it is doubtful already whether the same course or textbook can be used optimally with "anyone anywhere in the world," as Varner and Beamer [3, p. vii] put it. When you come to think of it, it is quite surprising that authors who discuss and analyze cultural differences at length believe that there is one product which is culture-neutral and can be used everywhere, i.e., their own course or textbook. Assuming that the *same* introduction will be optimally adapted to all audiences throughout the world is an example of cultural bias and should be avoided.

Previous Exposure of the Students

A course or textbook on any topic must take into account the level of knowledge that the audience already has, and determine on the basis of that from where to start.

There is little doubt that, on the average, American students will have experienced less exposure to foreign cultures when they come to your intercultural communication class than Europeans. There are, of course, simple reasons for this. In the United States, it is possible to travel long distances while remaining within the national borders. Without underestimating the cultural differences between different regions of the United States, students who travel there are unlikely to experience cultural differences on the same scale as in Europe. They may travel for days without having to cope with another language, a different

[1] I emphasize: *some* Asian cultures, probably not all.

currency, totally different opening hours for shops and stores, very different assortments of food in the supermarkets and restaurants, differences in the dominant ways of payment,[2] different driving habits, and so forth. In Europe, on the other hand, many students will have been exposed to these and other differences many times a year before they attend your class.

Culture shock is a universal problem, for no one can live in a foreign culture and remain unscathed; but Americans are likely to experience more severe culture shock than Europeans. Estimates concerning the rate of premature return of American expatriates run sometimes as high as 70 percent, and are often above 50 percent. For European (and Japanese) companies, these rates are consistently much lower, often in the 10 to 15 percent range.

American students who take my intercultural communication class in Maastricht often appear extremely surprised at aspects of everyday life in Holland that may seem trivial to Europeans: the fact that you seat yourself in many restaurants, that you don't get a glass of water with your meal, that you normally don't take home the leftovers, and so forth. Typically, American universities who send a number of their students over to Europe or to other places in the world will pay for a resident American staff or faculty member to live in the students' surroundings, and one of his/her tasks will be to soften the multiple culture shocks. European students who study abroad are not accompanied by such a faculty member from their home institution, and probably would not want to be.

For these and other reasons, a course in intercultural communication that is geared toward American students may have to start from a more basic level than it would in some other countries. Victor's *International Business Communication* [1] contains a map of world climate regions, of linguistic boundaries in Switzerland and Spain, of world religions, etc. This is no doubt useful in a country such as the United States where students rank low in their understanding of foreign cultures;[3] it may be less needed in a European setting (even if we should not overestimate the geographical knowledge of European students either!).

With respect to non-Western cultures, many Europeans and Americans tend to *under*estimate the amount of knowledge non-Westerners have about the West, and about the United States in particular. Many non-Western countries have been colonized in the past, and even those that have not are likely to have been exposed to Western habits and ways of doing things, particularly in business. Many of my African friends know and understand European culture (though they show no signs of adopting it) in a way in which I will never understand their native culture.

[2] As an example, checks account for 60.2 percent of all payments in France, as opposed to 9.7 percent in Germany; payments by credit card make up 12.2 percent of the total in Great Britain, as opposed to 1.2 percent in Germany (1989 figures, in [8]).

[3] In a UN study of 30,000 students age ten to fourteen in nine countries, U.S. students ranked next to last in their understanding of foreign cultures, and "one in every four high school seniors in Dallas, Texas, could not identify the country bordering the United States to the south" [2].

My Filipino and Thai students also have a fairly extensive knowledge of Western food, dress, music, movies, etc., and even of underlying cultural values such as care for the elderly (or the lack of it in the West), respect for parents and teachers (or the lack of it), etc. Their knowledge of Western values and practices is in any case vastly superior to what the average American or European student knows about Thai, Filipino, Bantu, or even Japanese culture.

On the other hand, in the course of a discussion about this topic that I had with my students in Bangkok in 1995, it appeared that not one of them was capable of mentioning any salient differences between U.S. and European culture; to them, Western culture was one big whole, and they did not recognize any variant of it.

In conclusion, the trainer must carefully assess the level of previous knowledge and exposure of his/her prospective students and should not simply assume that this level is almost the same everywhere. To do so would constitute another example of cultural bias.

Language Problems

Whenever a trainer teaches a class to students who do not share his/her native language, the language of instruction will be English in nearly all cases. This is so much a matter of course that often, the question is not even asked. Apart from France and Rumania where I have occasionally lectured in French, I cannot remember a single example where I used any language other than English, whether I was teaching or lecturing in the Philippines, Thailand, Poland, the Czech Republic, or other parts of the world. In quite a few cases English is the language of instruction I use even in Belgium, in particular in multinational companies and international institutions.

Typically therefore, both the trainer and the trainees will be using a language (English) that may not be their native tongue. Even if we assume that the trainer is capable of expressing himself/herself in English fluently (which is not always the case), the fact that the students have to listen and express themselves in English poses problems to a variable degree.

On the other hand, I am often surprised at the extremely high level of proficiency in English of many non-native students. This is, in my experience, particularly true in Northern and Central and Eastern Europe and to a lesser extent also in Southern Europe. But, as many scholars have pointed out, non-native speakers who are proficient in English may still be at loss when a native speaker uses uncommon words or expressions, including idiomatic expressions that have their origin in culture-specific fields such as (for the United States) baseball, basketball, and American football.

For reasons that are specific to the country and its history, the level of proficiency in English is near-native in the Philippines, where English is the normal language of instruction at the universities. In other Asian countries, however, the trainer should take into account that students' proficiency in English is uneven at best. In countries such as Thailand and Japan, well-known rules for communicating with non-native speakers must be followed very strictly:

- speak very slowly,
- repeat the information, preferably using different ways of expressing the same things,
- avoid difficult or rare words, idiomatic expressions, etc.,
- keep your syntax as simple as possible, avoid subordinate clauses and long sentences, and
- allow for frequent breaks: students tire more rapidly because they need to concentrate their utmost to follow the teacher in English.

In addition, in many Asian countries testing for understanding in the same way as in the West will be difficult, if not impossible. It is quite useless, in Japan, Thailand, and the like, to ask students if they understand, because to that question they will almost invariably reply "yes." Testing their understanding through more probing questions, on the other hand, may cause embarrassment. During breaks, however, conversation with the students may yield some insight as to the degree to which English is an obstacle to their understanding or not.

Emphasizing Only Certain Situations and/or Parts of the World

In the existing literature about intercultural communication in business, some parts of the world are discussed in detail, while others are not. There is plenty of material about Americans doing business abroad, and it is not difficult either to find material about cultural differences between the various Western European countries. Three of the main underrepresented fields, in my opinion, are the following.

- *The entire African continent.* This reflects the very limited role Africa plays in the global economy. Nevertheless, for some European countries, including France and Belgium, certain African countries (often the former colonies) are important partners both politically and economically (and for France also geostrategically) in ways that may not be reflected in bare trade figures. In providing intercultural training to (some) European audiences, therefore, the trainer should attempt to include more African examples and case studies in his/her training materials.
- *Intra-Asian encounters.* There is plenty of information in the literature about encounters between Asians and Westerners, and a Western trainer is likely to spend a high proportion of time in class discussing how *for Westerners* (Americans and/or Europeans) to do business in Asia. Not surprisingly, Japan

will be the country most often quoted, but it is also possible to find material on many other Asian countries such as Korea, Hong Kong, Singapore, Taiwan, China, Indonesia, Malaysia, or Thailand (see, for instance, [9]). However, intercultural encounters *between* Asians receive little or no attention: which are the main intercultural problems that arise when the Japanese do business in China, Korea, or Thailand for instance? This puts many Western trainers who teach (occasionally or permanently) in Asian countries in an awkward position, which some students may even feel to be tainted by ethnocentrism: most of the emphasis in the intercultural communication class will be on trade between the student's country and the West, while in reality the country's first trade partner may be Japan, and intra-Asian trade becomes more important generally. There is an urgent need for more material and analysis of intercultural issues between Asian countries.

- *Central and Eastern Europe.* Little material is available about the former socialist states in Central and Eastern Europe (CEE), and, therefore, the trainer will probably pay little attention to these countries in his/her class. For obvious reasons, any research carried out before 1989-1990 is unlikely to include the CEE countries. For instance, they are not part of Hofstede's survey [5, 10], while nearly all other parts of the world are represented in this major study. With the opening up of the borders and the economies of the CEE countries after 1990, however, and given that the expertise is not (yet) available locally, Western trainers are frequently invited to lecture in CEE countries. When I was asked to provide intercultural training in Gdansk, Poland in 1995, for instance, I could offer the trainees numerous examples of intercultural incidents and case studies from various parts of the world, but I had distressingly little material about Polish business culture and even less about potential problems that might arise when Poles do business with people from other cultures. There is a pressing need for intercultural research and publications concerning this part of the world.

Depth of the Analysis Provided in the Teacher's Course

A distinction is often made between a more superficial approach to the study of intercultural communication and a more in-depth analysis that offers higher explanatory power and real understanding. The most extreme form of superficiality is often considered to be a list of blunders or "do's and don'ts" without explanations, and many authors claim that their own material definitely offers more than that:

> A whole body of literature has appeared that documents blunders in intercultural business communication. Along with the horrifying or amusing stories are lists of do's and don'ts for businesspeople, as if remembering not to cross your legs in Thailand or not to refuse a cup of coffee in Saudi Arabia is all you

> need to know to close a deal. But lists of do's and don'ts don't tell you why you shouldn't cross your legs or say no to the coffee. And unless you understand the why of it, you may sooner or later fall on your face. The blunders-and-bloops literature is full of instances where the fall was fatal and the deal came apart. It is always because someone didn't understand the *why* [3, pp. 1-2].

I am not sure whether the cultural background of the trainer influences his/her tendency to favor in-depth analysis over a more superficial approach or the other way round. It is sometimes said that American trainers tend to simplify things (a caricature would be something like: "three steps to become a fluent intercultural communicator"), whereas Europeans would show their students how complex and difficult everything is. In reality, however, superficial-anecdotal books about cultural differences are commonly published in Europe as well as in the United States. Two well-known European examples are John Mole's *Mind Your Manners* [11] and Richard Hill's *EuroManagers and Martians* [12].

I would not reject all such publications wholesale. They are of widely varying quality. Some are little more than a mixture of information that seems to come straight out of an encyclopedia and vague, unsubstantiated generalizations about countries and people; they truly offer very little insight and understanding of cultural issues. A good case in point is Johnson and Moran's *Cultural Guide to Doing Business in Europe* [13], where we "learn" such things as: Austrians are "well known for their *Gemutlichkeit,* a happy approach to life" [13, p. 1]; "A Belgian typically puts great effort into both work and leisure" and "Leisure means family oriented pursuits—hobbies, sports, travel, books, . . ." [13, p. 9]; ". . . More than most peoples of Europe, the French abound in contradictions" [13, p. 49]; "Greeks enjoy life" [13, p. 75]; and so forth. In my view, such extreme generalizations are interchangeable and therefore meaningless. Couldn't we equally well say that the French enjoy life, the Belgians abound in contradictions, the Austrians put great effort into both work and leisure, and so forth? In addition, Johnson and Moran's book is full of errors, ranging from silly (*Deutsch* is said to be a Latin word [13, p. 62]), to gross (Belgium is said to have "highly productive natural gas fields [which] have made the country Europe's leader in gas output and exports" [13, p. 13]; in fact, the authors mix up Belgium and the Netherlands—hardly reassuring for intercultural specialists!).

Some other works, however, while descriptive rather than explanatory, are extremely well documented and reliable. They are valuable tools for anyone who is visiting the countries they describe. Engholm's *When Business East Meets Business West* [9] describes protocol, manners, and practices in several Asian countries, and does not concern itself much with underlying values or explanations. Yet in my view this book should be compulsory reading for anyone who is visiting any of the countries described.

Rather than rejecting the more descriptive approach offhand and claiming that an in-depth analysis is always preferable, I would suggest that the level of analysis depends on the audience the trainer faces. At least two factors may influence the type of training that is appropriate.

- *The kind of intercultural encounter the trainees are likely to face.* One does not train future expatriates who are going to spend several years in a foreign country in the same way as businesspeople who go to a country for a few days to negotiate a deal, then will be off to a different country one week later. My own activity includes training check-in and boarding personnel for an airline company, i.e., employees who, at check-in, spend two to three minutes with each client, and may face clients from dozens of different cultures within a few hours. In such a case, it would be a mistake (as well as nearly impossible) for the trainer to emphasize deep-seated value systems and underlying explanations over more superficial practices and manners related to different cultures.
- *The time available for the training.* There are thousands of different cultures in the world, and even if the trainer limits himself/herself to those that are most relevant in the global economy, it would not be difficult to fill 100 hours, or 100 days, with useful information about intercultural communication. But providing intercultural training to its employees is only one of the many challenges a company or an institution faces. Typically, time allotted to intercultural training programs is extremely limited; in my experience, usually about six hours for any individual working for a company. Sometimes, requests for a one-hour, or twenty-minute training session on "intercultural communication" are made! At universities, also, the course in intercultural communication competes with other priorities, and time is not unlimited.

In conclusion, the depth of the analysis provided in intercultural training programs will depend on many factors, including those I mentioned above. It cannot be said that one approach to training programs, such as the explanatory one, is in all circumstances preferable to another type, such as a more descriptive one.

Students' Responses

Asking Questions and Disagreeing

Student response will vary greatly depending on the culture. The importance of *face saving* in many Asian cultures is well known. Combined with high deference for the teacher, it entails that students in countries such as Japan and Thailand will seldom ask questions in class, let alone incisive or critical questions, and they are even less likely to point out mistakes that the teacher might make, even if these are simply the result of the teacher's absent-mindedness.

When discussing this topic with my students in Bangkok, I asked them what they would do if a teacher wrote an obvious error on the blackboard that could easily be corrected: would they point it out to him/her? The unanimous reply was: no, they would not. When I further asked whether there wasn't, then, any way in which they would inform the teacher about the error he/she made, one of the students replied with a wonderful expression: *our face would show a question mark.* It is very doubtful, however, whether a Western instructor would be able to read this indirect message and detect the question mark on his/her Thai student's face!

In my experience, *breaks* can play an essential role when teaching a class in Asian countries such as Thailand. Frequently students will walk up to their instructor during the break and, in private, ask questions and make remarks that they would never dare ask in front of the others, for fear of losing face themselves (if the question happened to be inappropriate) and/or for fear of causing the teacher to lose face (if the question implies even the faintest hint of criticism). It is advisable, therefore, to allow for more frequent breaks when teaching a class in certain Asian cultures than what one would usually allow for in the West (this is recommended because of the language problems also; see above).

Socializing after work has been described as a mechanism through which bonds between workers are created in Japan. In my own experience, such socializing also goes on between professors and students in Thailand. "Parties" are organized, often on the university premises, where students and one or more professors get together in the evening. One of the many occasions for such a party might be to honor those students who helped organize the orientation session for incoming freshmen or first-year graduates. Every student involved will receive a "certificate of gratitude," possibly with a token gift, he/she will give a thank-you speech, and so forth. On the same occasion, foreign guests such as visiting professors may also be honored, asked to give a short impromptu speech, etc. Dancing does not take place at such "parties,"[4] but one crucial socializing activity commonly *does* happen: karaoke singing, involving both professors and students. The foreign guest is well advised to participate in this activity! The karaoke singing creates a bond between the instructor and his/her students, and during such an evening, informal conversations with the students might reveal aspects and problems which would never surface in the more formal class setting.

Active Student Participation and Role Plays

Many intercultural training programs devised in Western countries include discussions of intercultural incidents and case studies, and possibly also role plays

[4] There is little dancing at parties in Thailand generally, even at weddings. My African friends and students, on the other hand, find it hard to accept that a party *without* dancing (as there are many in Europe) is worth the name.

and simulations. As I already mentioned, however, active student participation cannot be expected in the same way and to the same extent in different parts of the world. In many Asian countries in particular, when discussing incidents and cases, students may be preoccupied with various considerations such as the following:

- the desire not to seem better or more outspoken than the rest of the class,
- the desire to avoid emitting negative criticism about any features in the story discussed, and
- the desire not to contradict anything the teacher might say or (be believed to) think.

Getting trainees involved in role plays and simulations may be even more difficult. American audiences are usually quite willing to participate in such activities. In Europe, however, many people are ill at ease when performing role plays and will feel ridiculous in doing so. In addition, in some European countries with higher power distance, setting up role plays may pose additional problems. In a large international organization in Brussels, I regularly set up intercultural training sessions that are attended together by people whose position in the organization's hierarchy differs widely, from administrative assistants to high-level officials. When discussing incidents and cases, it is obvious that some (though not all!) lower-ranking personnel find it difficult to express themselves freely and openly in front of higher-ranking people. Under the circumstances, I never took the risk of setting up role plays, for fear that both lower and higher-ranking workers would feel embarrassed in participating in these.

These considerations are even more relevant in some Asian countries. With my Thai audiences, I find it hard even to imagine requesting them to perform a role play, although some students might conceivably be comfortable with them. My feeling is that it would be particularly difficult with older people, say over thirty. In reality, I never dared try it in Bangkok even with younger students, for fear of a potential catastrophe . . . In addition, I would have to ask the students to perform the role play in English rather than in their native language (Thai in this case), and this would add another burden to the students' plight that we cannot underestimate (see above).

In the Philippines (in 1992), on the other hand, I did Daphne Jameson's *Intercultural Communication Simulation*[5] with a class of twenty-year-old students, and while they were manifestly less at ease with the idea than a comparable group of Americans would have been, things went reasonably well, in my view about as well as in most European countries. For cultural and also for purely linguistic reasons, Filipino students are less reluctant than most other Asians to participate in such activities.

[5] This stimulating simulation, which is also fairly easy to set up, has now been published [14].

Getting Personally Acquainted with the Students

In Asia, the (foreign) teacher's personal involvement with his/her students may not need to go beyond what is described above; but *African* students will often attempt to get to know their instructor on a much more personal basis. This may be done during the break, but, in my experience,[6] ***extended office hours*** are the best way to cater for the need African students may feel for personalized contacts with their instructor.

A difficult question in this respect will be how far one can go. On the one hand, it will be hard for African students to attend a class by an instructor who does not show any warmth, humanity, or personal involvement toward his/her students: it is somewhat like attending a class taught by a machine. How can a teacher be indifferent to his/her students' financial problems when their scholarship is discontinued, when their family back in Africa needs money, etc.?

On the other hand, the more the teacher gets personally involved with his/her students, the higher the likelihood that some of them will start asking for "favors" which are against university regulations (such as taking an exam a second time when the result was poor at the first try, or being "lenient" when grading a paper; there are always very good personal reasons for such requests).

Usually, however, students will understand that one can be very firm as far as the rules are concerned, while at the same time showing understanding and warmth to the students and genuine interest for their personal problems.

Evaluating the Teacher

Evaluating the teacher is common practice in the United States, and many trainers routinely hand out evaluation forms to their students at the end of a course. Student evaluations are felt to be an essential part in improving one's teaching and tailoring it optimally to the audience's needs and expectations.

European trainers who work for commercial companies often follow the same practice. Some European university professors, on the other hand, are reluctant to accept open evaluations by their students. At my own university in Antwerp, Belgium, students do fill out evaluation forms about teachers and classes, but these are never made public. Individual professors receive only their own evaluation, in an envelope marked "confidential," and only a very limited number of other people, such as the university president and the vice-chancellor, have access to this "confidential" information!

This stands in sharp contrast with the practice at the University of California, San Diego, where I did some of my graduate work: here student evaluations are published as a brochure and available for all to buy at the university bookstore.

[6] In fact, my own teaching experience is not in Africa, but with the (numerous) African students I teach at the College of Development Studies of the University of Antwerp, Belgium.

Ateneo de Manila University in the Philippines, where I taught a class in 1992, also allows for extensive student evaluation in a manner that is similar to many U.S. universities. In Thailand, to my knowledge, the practice does not exist, and given the prevailing attitude of deference of students toward their teachers (*ajarn* or "teacher" is one of the highest terms of honor one can use when talking about someone), it is unlikely that any evaluation of teachers by their students could be successful.[7]

In any case, if the university or the institute where the Western teacher performs his/her training program does not organize such a student evaluation, it is prudent not to organize one on your own without checking the appropriateness and advisability of the enterprise first.

The Western Perspective of the Intercultural Communication Class

Even if the trainer manages to overcome the different kinds of cultural bias I outlined above, it remains that, in a course in intercultural communication *in business,* in particular, it will be especially difficult to avoid an overall Western perspective, and it may not be either possible nor even advisable to attempt to do so.

After all, doing business the way it is done today is largely a Western invention, and while some other countries, in Asia in particular, may now outperform Western nations because of certain cultural traits they possess (such as long-term orientation), the way the economy and trade are organized on a global scale follows Western patterns.

I do not think it really possible to teach a course in intercultural communication *in business* without making many basic assumptions such as, for instance, the following.

- Trainees wish to be economically successful and have no desire to detach themselves completely from material possessions (the latter might be the philosophical position of orthodox buddhists, even if many buddhists attempt to reconcile the Buddha's teachings with the pursuit of economic success).
- The money one earns when doing business should be at least in part reinvested or saved, rather than distributed entirely to relatives who might be in need (the latter could be the prevailing attitude of many Africans).
- When certain features of a modern economy contradict traditional moral and/or religious precepts, a trade-off or compromise between both should be sought, rather than giving absolute priority to the traditional precepts (the latter might be the position of conservative Muslims).

[7] The newly appointed American manager of a large hotel in Bangkok almost created a row when he wanted all workers to evaluate their superiors; the scheme never materialized [7].

Even if these and other basic assumptions cannot be thrown overboard without threatening the very function of a course in intercultural communication in business, the least we can do is to be *aware* of our own culturally defined ideology in this respect.

CONCLUSION

In this chapter I argued that no one is immune to cultural bias, and that intercultural trainers must critically examine their methods, course materials, and teaching aids for possible instances of it (Cultural Bias, p. 192).

I then presented a list of factors which may lead to cultural bias. This list can be used as a checklist by anyone who is attempting to detect and eliminate instances of cultural bias in his/her own teaching (Cultural Factors that may Influence the Trainer-Trainee Relation in Intercultural Communication Classes, p. 196).

I expressed serious doubts whether it is possible, or even desirable, to make intercultural training materials and instructors culture-neutral, just as it is difficult to make communication patterns or commercial products culture-neutral. A good intercultural trainer must, however, to the highest possible degree, be *aware* of his/her own cultural biases.

> We need to be aware of our own biases. We need to be open about the discovery within ourselves of unrecognized biases that can exert an influence on how we understand another culture [3, p. 16].

Only then can one consciously decide which instances of cultural bias seem unacceptable, and which ones are so basic to the teacher, the objectives of his/her course and the interaction with the students that they must be maintained. When this is made clear explicitly during class, most students will find it acceptable that a course in intercultural communication *in business* starts from certain premises that cannot be questioned without making the course itself impossible or meaningless.

REFERENCES

1. D. A. Victor, *International Business Communication,* HarperCollins, New York, 1992.
2. G. P. Ferraro, *The Cultural Dimension of International Business,* Prentice Hall, Englewood Cliffs, 1994.
3. I. Varner and L. Beamer, *Intercultural Communication in the Global Workplace,* Irwin, Chicago, 1995.
4. N. Adler, *International Dimensions of Organizational Behavior* (2nd Edition), Wadsworth, Belmont, California, 1992.
5. G. Hofstede, *Culture's Consequences. International Differences in Work-Related Values,* Sage, Beverly Hills, California, 1980.
6. R. Pernoud, *Pour en finir avec le Moyen Age,* Seuil, Paris, 1977.

7. S. P. Verluyten, *A Course in Intercultural Communication in Business and Institutions*, ACCO, Leuven, Belgium, 1996.
8. Kredietbank, *Bulletin hebdomadaire, 46*:20, pp. 1-6, 1991.
9. C. Engholm, *When Business East Meets Business West. The Guide to Practice and Protocol in the Pacific Rim*, Wiley & Sons, New York, 1991.
10. G. Hofstede, *Cultures and Organizations. Software of the Mind*, McGraw-Hill, London, 1991.
11. J. Mole, *Mind Your Manners*, Industrial Society Press, London, 1990.
12. R. Hill, *EuroManagers and Martians. The Business Cultures of Europe's Trading Nations*, Europublications, Brussels, 1994.
13. M. Johnson and R. T. Moran, *Cultural Guide to Doing Business in Europe* (2nd Edition), Butterworth-Heinemann, Oxford, 1992.
14. D. Jameson, Using a Simulation to Teach Intercultural Communication in Business Communication Courses, *The Bulletin, LVI*:1, pp. 3-11, 1993.

CHAPTER 10

Culture and the Shape of Rhetoric: Protocols of International Document Design

ELIZABETH TEBEAUX
and LINDA DRISKILL

Changes within the global village have posed a major challenge for business communication faculty. What methods can be used to prepare students efficiently and effectively for communicating in cross-cultural business settings? Clearly, no faculty member can possess competence in every culture. Within the constraints of the business communication class, faculty have limited time to devote to international communication strategies. Even in an international business communication course, only selected countries and cultures can be addressed. To attempt to address these pedagogical issues, we wish to suggest a method that will achieve four goals. Our approach:

1. introduces students to some of the cultural variables ("values") that influence document design,
2. illustrates how specific values translate into design choices in different cultures,
3. suggests questions that students must answer before designing documents for readers in specific cultures, and
4. familiarizes students with the demands of preparing business communications for readers in other cultures in an efficient way.

MAJOR ISSUES IN INTERNATIONAL BUSINESS COMMUNICATION INSTRUCTION

As Connor noted in her studies of contrastive rhetoric, research in written business communication is sparse [1, 2]. Much of the problem lies in the difficulty of faculty access to primary materials. However, another issue is the problem of change spurred by technology. Charles Kostelnick has shown that new technologies create complex changes in communication designs [3]. Which countries' aesthetic criteria will dominate the design of printed documents and screens on the World Wide Web when business people from different countries communicate with one another [4]? What approach should communicators follow in designing documents for audiences across borders [5]?

The answers are complicated. First, the acquisition of new technologies is not uniform: some countries and companies adopt them much more readily than others. Second, although the firms creating electronic technologies, especially software, usually follow U.S. notions of design, greater nationalism everywhere tends to valorize local traditions and increases resistance to global standards. Finally, processes of technology adoption vary, and it is not clear whether multinational companies will adopt one cultural aesthetic and rhetorical norm throughout a firm or adapt to local norms in the countries where they do business, even if they use other norms for communicating with headquarters [4].

However, one fact is clear: Greater interaction and differences in cultural preferences increase the probability of miscommunication, including pragmalinguistic and sociopragmatic failures [6]. Writers feel uncertain about how to plan communication and how to compose. Because people make judgments about one another's characters, intentions, and attitudes on the basis of document characteristics, we believe document design—page design, layout, content, and style—will remain an important issue. As corporations adopt new electronic communication technology, we expect changes in the rhetoric and the appearance of both hard-copy communication and electronic communication, but most of us are apprehensive about what these changes will be and what we should teach students and business people.

This chapter offers a way to approach document design during a period of rapid change. The scheme of contrasting U.S. and non-U.S. cultural preferences that we propose is, quite frankly, an enormous oversimplification. We suggest it simply as a way of challenging students' cultural complacency and as a starting point for their developing more elaborate protocols that take into account many distinctive national cultures and complex multicultural audiences. Moving from the classroom to the office, students should be prepared to move beyond their initial exposure to both national and international communication strategies. Furthermore, what we propose should enable students to see that corporate as well as national cultures interact in a single context. However, we believe that understanding basic differences between U.S. culture and other cultures must underpin students' approach to the development of international written business communications. Because no teacher can prepare students to write documents for readers in more than a few cultures, we do not propose to explain how to design a document for a specific audience, but to demonstrate to students that the U.S. model of document design is not universal and that design of documents is culturally determined.

DEFICIENCIES IN CURRENT DOCUMENT DESIGN RESEARCH AND COMMUNICATION SOFTWARE

As we said, the problem today is not lack of interest in international communication research. Since 1986, increasing numbers of articles and monographs have been published on communications problems and strategies for multinational organizations. These studies focus on managerial communications—dealing with diverse work forces, recognizing cultural differences among employees, planning and controlling oral communication with multicultural employees [7]. Only a few studies have explored methods of preparing documents—letters, memos, e-mail, and reports—for international readers, aside from reference works on business correspondence protocols (salutations, addresses, and closings) [8] and articles on software documentation [9-11]. A few are devoted to genres in a particular country, such as Bell, Dillon, and Becker's article on German memo and letter style [12] and Varner's comparison of French and American correspondence [13], Yli-Jokipii's work that compares British, Finnish, and American business writing [14], Jenkins and Hinds' study of English, French, and Japanese business letters [15], and Connor's comparison of American and Japanese business correspondence [1].

A number of excellent texts warn writers about the problems arising from culture and ways it shapes readers' perspectives [16-18]. Hoft's textbook focuses on international user analysis [19]. A number of studies caution writers to be aware of how the meanings of words and graphics may change across cultures [21-24], and most support this thesis with examples from different cultures. Despite the increasing number of studies on business writing in various cultures, the problem remains that few guidelines exist for helping writers develop routine written business documents, such as correspondence, reports, and mail that will be written in English for use by readers in other cultures. Even fewer take into account visual and graphic conventions.

REDEFINING THE MEANING OF DOCUMENT DESIGN

The special 1989 *Technical Communication* issue on document design emphasized the range of activities involved in creating communication that is simultaneously visually and verbally effective. In the lead article, issue editor Karen Schriver defines document design as "the theory and practice of creating comprehensible, usable, and persuasive text" [25, p. 1]. She notes the process of document design is not merely formatting text to make it visually appealing but is more like information design. The process requires thorough analysis of the audience's needs as well as decisions about content, style, and layout to yield the most easily used and persuasive communication.

Attention to the culturally defined context of a document has been missing in recent document design studies. Document design formerly was thought of as the more technical design process of preparing a document for print—layout, paper choice, colors, and so on. However, a substantial number of communication researchers in the past fifteen years have emphasized the interaction of these design elements with rhetorical elements. They conclude that verbal and visual elements together cause readers to recognize a genre, construct an intended message, and relate it to a particular situation [25-37]. These researchers have mainly focused on U.S. document design practices without explicit concern for other cultures. Instruction in international business communication must not rely on the results of these studies because not all cultures value efficiency or clarity in the way that is implied by these studies or require all communication choices to have the same tightly focused cohesiveness. In short, U.S. definitions of document design overlook the emphasis that non-U.S. cultures may place on other types of communication functions and goals. A consideration of how U.S. business culture differs from others will reveal differences in purposes, functions, and visual as well as verbal conventions of written communication.

Why a New Definition of Document Design is Essential

In a broad sense, the U.S. business culture contrasts significantly with other business cultures throughout the world. United States business culture prizes order, efficiency, directness, and effectiveness. Viewed from a U.S. standpoint, the ideal document design is an efficient system of well-defined, clearly interrelated elements, all of which contribute appropriately to the dominant message and guide the eye of the beholder. These values are clearly implicit in popular U.S. document design books such as Roger Parker's *Looking Good in Print* [38] and *The Makeover Book* [39]. Better designs are said to eliminate clutter, unnecessary words, and excessive visual elements. They lead the eye through contrasting text and graphic hierarchies and use white space skillfully to guide the reader's eye in a desired sequence for efficient detection of information. In contrast, other cultures prefer, in differing degrees, a design in which the status of verbal and visual elements depends less on coherence of the document and more on elements' relation to external reference frames. Signals conveying abundance and status or inviting aesthetic pleasure—even digressions—may be considered appropriate. Efficiency and coherence may not be considered necessary from this perspective and may even be seen as crass or rude. Embellishments and design choices related to aesthetic, social, or ethical values—even religious values—may be seen as more desirable.

Furthermore, in defining document design as a process of choice of layout, organization, content, and style, theorists and researchers have assumed that writers know what choices are possible. Writers' perceptions of available choices will be biased by their own culture's preferences and restricted by their contact with other cultures. To begin to provide a strategy for dealing with these limitations, we wish to address how a communicator can discover the wider range of appropriate choices and to provide guidelines for document design. These guidelines reflect changes in communication technology and cultural perspectives.

United States produced software programs for designing printed and electronic documents are not likely to be sensitive to non-U.S. readers' cultural preferences. Software features and templates are likely to reflect the preferences of the high-tech cultures in which they were produced. Easy to use (but difficult to customize) software is likely to swamp a new user's cultural traditions and preferences. For example, early versions of Adobe Pagemaker™ (formerly Aldus) did not allow an image to be rotated and presented diagonally; persons from a culture that valued such a design characteristic would have been helpless to achieve it. Decisions about many software design features should rest on excellent research of multiple cultures.

Weiss [40] and Driskill [5] conclude that we should not emphasize the otherness of audiences from other countries. Emphasizing awareness of cultural dimensions and possible combinations of design choices provides a starting point for developing a contextually appropriate document design that expresses maximum cultural respect and harmony, not slavish adherence to stereotypes. It is convenient to examine, in broad brush outline, how U.S. culture compares to non-Western cultures. From such an examination, we can observe the opportunities for adaptation when U.S. writers design written documents for non-Western readers. Observing major differences in selected documents available to the faculty member will help students anticipate points of decision when designing for non-U.S. readers. A knowledge of general conventions must still be modified by consideration of the company cultures and the departmental cultures that also affect how readers perceive a particular document. The six value dimensions discussed later in this chapter also help readers predict and understand differences in document design in the texts they receive from other countries.

Document Design: The Need to Reflect Cultural Value Dimensions

Recent research on international business falls into two categories, but neither one links culture to written communication practices [18, p. 5]. One approach assumes that certain national cultures share common values and can be grouped into value and behavior classifications. A second approach focuses on a single country and its cultural features. This second approach is illustrated by the increasing number of books on appropriate behavior for doing business in specific countries—e.g., Doing Business in Mexico, Doing Business in Japan, and so on.

A thorough mastery of a culture, as it influences written communication, would be ideal; but, as we have mentioned earlier, most communicators lack the necessary expertise to prepare business letters and reports for readers in another culture. United States business communication students also need a foundation on which they can begin to build expertise in designing written international reports and letters. As Shuter and Wiseman [18, p. 6] observed, the majority of these country-specific studies have focused on Japan and the Peoples Republic of China [41]. To begin to fill the gap in instructional methods, we propose that instructors 1) present several value dimensions identified by Hofstede [42-44], Trompenaars [45], Hall and Hall [21], and derivative studies, 2) explain how U.S. business culture differs from other cultures in terms of these value dimensions, and 3) base class assignments on the implications of these dimensions.

Six cultural dimensions that seem to have great importance for international document design heuristics are:

- value of (emphasis on) either individuals or groups
- either the separation or the merging of business and private relationships
- degree of distance between social ranks
- universal or relative (particular, situated) view of truth
- expectation either that a message will include all necessary background information or that reader's knowledge of matters not included in a document will make correct interpretation possible
- predisposition either to accept or to avoid uncertainty.

To illustrate how these dimensions affect document design, we will describe several examples and discuss their style and development before concluding with observations on basic pedagogy for introducing students to international business communication.

UNDERSTANDING DIFFERENCES IN WESTERN AND NON-WESTERN CULTURES: SAMPLE DOCUMENTS

The examples in this study consist of routine business documents from Pakistan as well as documents from a special situation described in 1993 U.S. Congressional hearings held before the passage of NAFTA in 1994 [46, 47]. These hearings focused on abuses within the Mexican political system. The documents, included as appendices to the transcript of the hearing, are readily available and illustrate the application of many of the concepts we have discussed.

We will also include an additional Mexican business letter because it illustrates other products of cultural values stressed by Hofstede.

Routine Documents from Pakistan

Two routine documents from Pakistan, a collection letter and a request for a bid and price quote, illustrate contrasts in acceptance of U.S. formats, acceptance of power differences, merging of business and personal relationships, low uncertainty avoidance, full textual embodiment of the message, and a universal view of truth, values that will be discussed later. The Gannon Dunkerley collection letter (Figure 1) uses the greeting "Dear Sir," although the writer obviously knows the intended reader's name, which is shown on the attention line as "for the kind attention of Mr. <Specific Name>." The tone is polite but formal and indirectly expresses the request: "We draw your kind attention to the long pending Bills . . ." The letter format is organized around a central axis with an elaborate centered letterhead. The closing and signature after the company name are positioned British style, on the right margin rather than the left.

In Figure 2, the Northern Traders telefax message, the trading company acknowledges that one of the reader's clients, Sitara Chemicals, has referred Northern Traders to this supplier. According to Mr. Anjum Bilgrami, a Pakistani engineer, the referral is necessary to justify the reader's considering a relationship with a new buyer. In the telefax letter, Northern Traders is seeking bids for an Italian firm. Northern Traders also does business with clients from many other countries, such as China, and handles the sales of complete plants as well as the marketing of specific manufactured products. Although it follows the Pakistani cultural preference for identifying the business referral, in other aspects of its correspondence it reveals an adaptation to up-to-date technology, U.S.-style unindented [block] paragraphs, and left-margin alignment of the greeting and close. The typist somewhat unpredictably capitalizes "important words," rather like an eighteenth-century British document, but also uses modern listing technique.

A Non-Routine Situation: The U.S.-Mexican Scenario

We wish to describe briefly the dilemma faced by Kaveh Moussavi, a Mexican citizen working as a commercial consultant in London, who, in 1992 was appointed by IBM to help win a contract for air traffic control equipment in Mexico. In delivering IBM's bid, Mr. Moussavi had been asked to contribute $1 million to the Salinas government in exchange for IBM's receiving consideration in the bidding process. Mr. Moussavi at first had believed that he had been approached by individuals impersonating government officials (they showed no credentials and did not give their names). He had reported what he thought were people who would discredit President Salinas' reform programs. However, after Mr. Moussavi reported to the Mexican government that he had been approached, he was publicly denounced by the Mexican government and the Mexican press;

TELEGRAMS : "GANNON"
CABLES : Karachi

PHONE : 5681190
Telex : 29717 GANON PK
Fax : 00 92 21 568 4840

GANNON DUNKERLEY & CO. PAKISTAN (PVT.) LTD.

Merchants, Engineers & Contractors

REGISTERED OFFICE :
PAKISTAN INSTITUTE OF INTERNATIONAL AFFAIRS BLDG.
1, AIWAN-E-SADDAR ROAD, KARACHI-74200.

OUR REF H/3076/SM-74
YOUR REF ....

IMPORT REGISTRATION NO. K04360
EXPORT REGISTRATION NO. W004786

G. P. O. BOX NO. 460
KARACHI-74200 (PAKISTAN)

Date.... Nov. 6, 1995.

The Chief Executive,
Name Sugar Mills Ltd.,
Lakson Square Building,
K a r a c h i.

for the kind attention of
Mr. S.H. Name

Dear Sir,

RE: PENDING BILLS

We draw your kind attention to the long pending Bills :

1)	Bill No.	ENG/SM/817/7	dt. 8.8. 94	Rs.50,000.00
2)	" "	TM/TC/773/42	" 30.11.94	Rs.32,000.00
3)	" "	TM/GEN/771/64	" 30.01.95	Rs. 1,750.00
				Rs.83,750.00

During our visit last month, your Mr. S.H. Name very kindly promised to settle the above Bills by the first week of Nov. The Bills are pending for more than a year and in addition we are in urgent need of funds. We, therefore, request you to please send us your Cheque in settlement of our above Bills soon after receipt of this letter.

Your co-operation is solicited.

faithfully yours,
for GANNON DUNKERLEY & CO. PAKISTAN (PVT) LTD.

FIRSTNAME LASTNAME
DIRECTOR

Figure 1.

NORTHERN TRADERS (PVT) LIMITED
17-J, BLOCK-6, P.E.C.H.S.
KARACHI-75400
FAX NO : 92-21-454-61-20, TELEX NO : 24364 HUSGR PK

TELEFAX MESSAGE

COMPANY NAME : ( ) - ITALY NO. OF PAGE 2
(INCLUDING COVER PAGE)

FAX REF. : F-1045/94
FAX NO. : ( )

ATTENTION : MR. F.LADAVAS <DIRECTOR COMMERCIAL> DATE :24.7.94

Dear Mr. Ladavas,

Mr. A. Sattar of Sitara Chemicals has given me your name and Fax No.

Our Company

M/S NORTHERN TRADERS (PVT.) LTD.
17 - J, BLOCK - 6, P.E.C.H.S.,
KARACHI - 75400. PAKISTAN
FAX - (021) 4546120.
TLX - 24364 HUSGR PK.

Is engaged in the sales of Complete Plants to the Pakistan Market. Currently we are working with a client for a Complete Plant which requires about 15000 Tons/Annum, 100% Basis, Caustic Soda as a liquid at 32% concentration. Caustic Soda is a basic Raw Material for the Plant.

We would appreciate a complete offer for the following .

a. PRODUCT :
Caustic Soda.

b. CAPACITY :
15000 TPA on 100% Basis as a 32% Liquid.

c. SCOPE OF SUPPLY :
Complete plant within battery limits.

Your offer should comprise two parts as follows:-

a. TECHNICAL OFFER :
1. Process Description.
2. List of Machinery.
3. Raw Material Specs.
4. Product Specs.
5. Per Ton - Product, Raw Material Consumption Figures.
6. Per Ton - Product, Utility Consumption Figures.
7. Layout.

b. COMMERCIAL OFFER :
1. FOB price for Machinery, Design and Engineering.
2. Proposed Terms of Suppliers Credit.
3. Delivery period.
4. Prices in b.1. should be net. We will add Commission at our end.

We hope you have understood our requirement and will let us have your offer at the earliest.

Best Regards.

Figure 2.

his family, living in England, was threatened; and his own life and reputation were in jeopardy. Mr. Moussavi decided that the persons asking for the bribe had, after all, been corrupt government officials. The documents presented at the hearing provide evidence of Mr. Moussavi's efforts to clear his name and expose the corruption of the Mexican government (Abuses [1993], 7-13). We provide the bulk of the transcript in Appendix A.

Figure 3 [46, pp. 352-353]. On April 6, 1993, Mr. Moussavi writes to the secretary of the ministry of transportation and communication. Paragraph 1 sets the tone of the letter. He uses consistent formality throughout the letter and appeals to his reader—Secretary E. Gamboa Patron—in terms of the specifics of the situation rather than the morality of the situation, a posture he later assumes in Figure 5. Note that he indirectly, rather than directly, alludes to the attacks on him. The style is fluid, formal, but moderately florid. He moves gracefully into his request, apparently because Mexico is a high-power distance, group-oriented culture that accepts uncertainty and views truth as related to the particular details of a situation (paragraph 2).

Figure 4 [46, p. 358]. This letter of June 18, 1993, addressed to the Mexican Consular General of London, is clearly reminiscent of a U.S. letter. Mr. Moussavi states his specific purpose in the opening paragram. Note that in addressing this individual, who is Mexican but resides in England, Moussavi is blunt and direct in his statement. He alludes to "this sorry affair" and the "hypocrisy of the Mexican government." In contrast to Figure 3, which uses an approach and a style that are indirect and syntactically loose, Figure 4 leaves no question about what Mr. Moussavi wants from his reader. The conventions of legal protest may explain Moussavi's decision to design his letter in this way. All Mr. Moussavi's correspondence with people in England (included in the transcript of the hearings) has this same directness and open statement of purpose.

Figure 5 [46, pp. 371-374]. This letter, addressed to President Salinas himself on June 25, 1993, illustrates a third sharp redirection in tone by Mr. Moussavi. Note the formality of the salutation and the opening paragraph, the indirect explanation of the events surrounding the bribe, the failure to attack the corruption of the Salinas regime (a point Mr. Moussavi makes repeatedly in the hearing and in other letters), and the shift to an effacing, indirect supplication to Salinas for his help in clearing Moussavi's name [46, p. 371].

Figure 6 [47, pp. 107-108]. This letter, written six months earlier in the case (January 11, 1993) by Carlos F. Poza of the American Consulate in Mexico to the Comptroller General of Mexico, is positive but indirect in its statement of the problem—that U.S. companies were not given a fair opportunity to win the bid. Mr. Poza states his concern about the situation and offers his assistance in investigating the complaints. Note the long sentences and the deliberate lack of directness in informing Mr. Jimenez of the problems that Mr. Moussavi has experienced. The letter exemplifies the typical Mexican business letter: indirect presentation, diffuse statement of facts, lack of conciseness, and gracious style.

352

Kaveh Moussavi
56, Old Road, Oxford, OX3 7LL,
United Kingdom
Tel + 44 865 742374 Fax + 44 865 750065

16 April 1993

His Excellency E. Gamboa Patron,
The Secretary of State for Transport and Communications,
Ministry of Transport and Communications,
Mexico D.F.
Mexico

Dear Mr Secretary,

First and foremost allow me to extend my sincere congratulations to your excellency on your appointment to your new post. You will, no doubt, be aware that I have a special reason for welcoming this change at the top of the Ministry of Transport. It is my sincere hope that your arrival at the helm may usher in an intense investigation into establishing the accuracy, or otherwise, of my statements published in the Financial Times on February 3rd.

Your excellency will, I am certain, not be unaware of the one sided, libellous and highly unfair propaganda drive unleashed in the Mexican press against me in the last two months. Your predecessor, within two days of the publication of my statements made an absurd, if not comical, statement to the press purporting to have carried out a full investigation into my allegations and had concluded that I was a liar and should be thrown into prison. Precisely how he could have carried out such a detailed investigation leading to such definite conclusions, it has never been explained. In any event Sr. Caso by taking such a forceful and public position so early on in the case, ended up boxing himself

353

into a position and thereby disqualified himself from presiding over anything like a serious investigation.

That is precisely why I specially welcome the appointment of your excellency as the head of the Ministry of Transport and Communications. Can I hope that you will, as a matter of greatest urgency, order an intense and unbiased investigation into the recent SENEAM tender? In doing so you can most certainly count on my total co-operation as has already been confirmed by my attorney in correspondence with your government.

I therefore request that you order your embassy in London to take a full statement from me. I think you will understand my reasons for not wishing to travel to Mexico in the present climate of witch hunt that has been instigated by the government controlled media. This is the only condition that I would ask you to accept. Otherwise I shall be available to assist with whatever line of enquiry that your investigators might wish to follow.

Excellency,

I have the honour to remain at your service

Figure 3.

368

Kaveh Moussavi

56, Old Road, Oxford, OX3 7LL,
Tel 0865 742374 Fax 0865 750065

18 June 1993

Martin Brito, Esq. ,
Consul General,
Consulate General of Mexico,
8, Halkin Street,
London SW1 7DW

Dear Mr Brito,

This is a note further to my facsimile of yesterday, in which I served notice that I wished to file a formal judicial complaint against Sr. Ibarrola of the Mexican Foreign Ministry. I did specifically request that you acknowledge receipt of that fax. I have also phoned and spoken to Minister Castro Valle. I tried to reach you on the phone and discovered to my amazement that, for the first time throughout this whole sorry affair, you were not available to speak to me!

Can I please request that;

1. You acknowledge immediately receipt of my fax of yesterday
2. Arrange and let me know the time of my appointment to come to the Mexican Consulate in order to file a formal complaint against Eduardo Ibarrola as stated in my letter of yesterday.

If I do not hear from you on Monday, I will take this as yet further confirmation of the hypocrisy of the Mexican government, and its persistent determination to prevent the truth about this whole affair being revealed to the public.

Yours sincerely,

Figure 4.

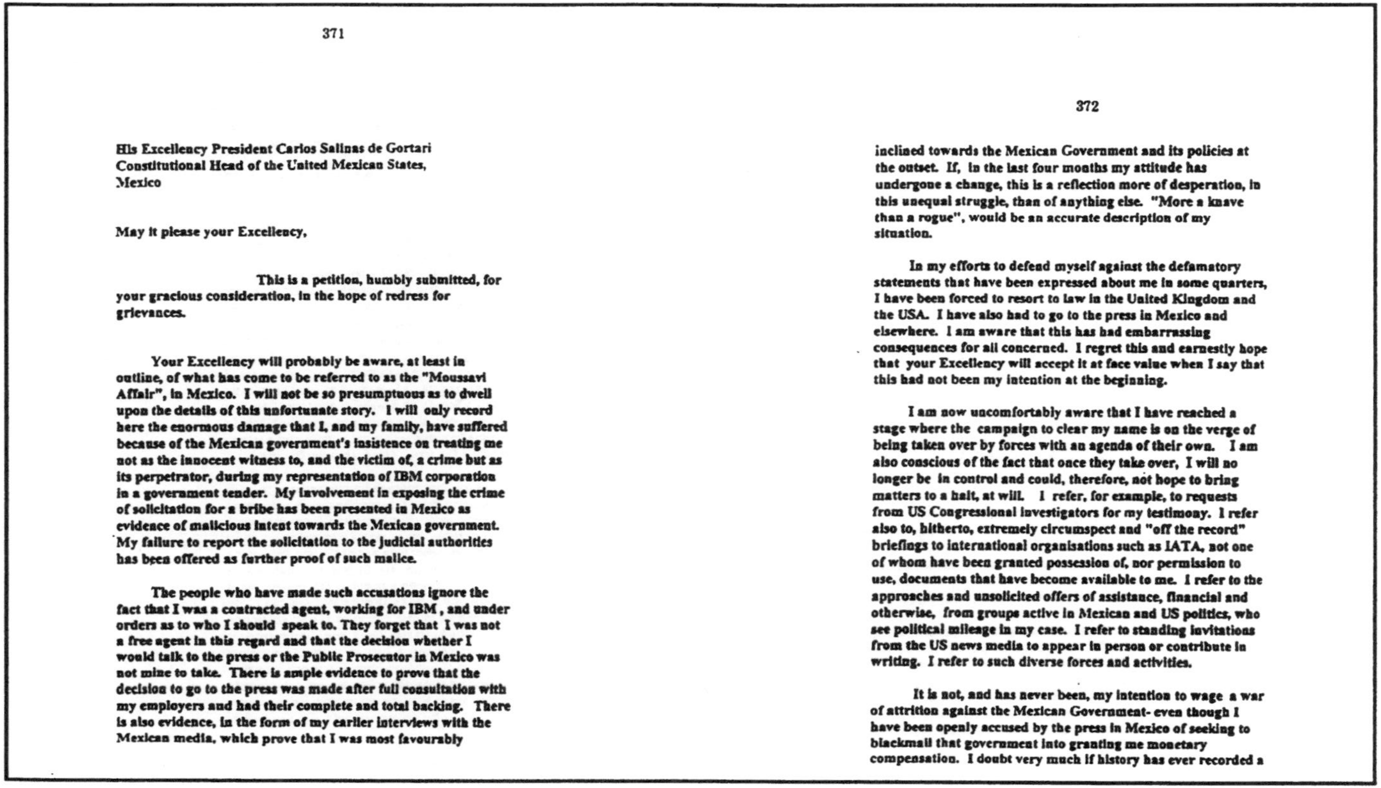

371

His Excellency President Carlos Salinas de Gortari
Constitutional Head of the United Mexican States,
Mexico

May it please your Excellency,

This is a petition, humbly submitted, for your gracious consideration, in the hope of redress for grievances.

Your Excellency will probably be aware, at least in outline, of what has come to be referred to as the "Moussavi Affair", in Mexico. I will not be so presumptuous as to dwell upon the details of this unfortunate story. I will only record here the enormous damage that I, and my family, have suffered because of the Mexican government's insistence on treating me not as the innocent witness to, and the victim of, a crime but as its perpetrator, during my representation of IBM corporation in a government tender. My involvement in exposing the crime of solicitation for a bribe has been presented in Mexico as evidence of malicious intent towards the Mexican government. My failure to report the solicitation to the judicial authorities has been offered as further proof of such malice.

The people who have made such accusations ignore the fact that I was a contracted agent, working for IBM, and under orders as to who I should speak to. They forget that I was not a free agent in this regard and that the decision whether I would talk to the press or the Public Prosecutor in Mexico was not mine to take. There is ample evidence to prove that the decision to go to the press was made after full consultation with my employers and had their complete and total backing. There is also evidence, in the form of my earlier interviews with the Mexican media, which prove that I was most favourably

372

inclined towards the Mexican Government and its policies at the outset. If, in the last four months my attitude has undergone a change, this is a reflection more of desperation, in this unequal struggle, than of anything else. "More a knave than a rogue", would be an accurate description of my situation.

In my efforts to defend myself against the defamatory statements that have been expressed about me in some quarters, I have been forced to resort to law in the United Kingdom and the USA. I have also had to go to the press in Mexico and elsewhere. I am aware that this has had embarrassing consequences for all concerned. I regret this and earnestly hope that your Excellency will accept it at face value when I say that this had not been my intention at the beginning.

I am now uncomfortably aware that I have reached a stage where the campaign to clear my name is on the verge of being taken over by forces with an agenda of their own. I am also conscious of the fact that once they take over, I will no longer be in control and could, therefore, not hope to bring matters to a halt, at will. I refer, for example, to requests from US Congressional investigators for my testimony. I refer also to, hitherto, extremely circumspect and "off the record" briefings to international organisations such as IATA, not one of whom have been granted possession of, nor permission to use, documents that have become available to me. I refer to the approaches and unsolicited offers of assistance, financial and otherwise, from groups active in Mexican and US politics, who see political mileage in my case. I refer to standing invitations from the US news media to appear in person or contribute in writing. I refer to such diverse forces and activities.

It is not, and has never been, my intention to wage a war of attrition against the Mexican Government- even though I have been openly accused by the press in Mexico of seeking to blackmail that government into granting me monetary compensation. I doubt very much if history has ever recorded a

Figure 5.

373

single case of a blackmailer hiring two top law firms, in the USA and the UK, to put his case to the intended victim. Nor, of course, is there a single recorded instance of my having ever asked for monetary compensation. For the truth is, your Excellency, that given only a fraction of a chance, I would seize it immediately and bring this campaign to a halt, while it is still in my power so to do. But in all honour, I am bound to say that I can not just drop everything and pretend that I have not been libelled, my name and business not destroyed, my family very nearly wrecked by the actions of Mexican officials. I can not simply disregard the enormous damage that has been inflicted on me as a result of the Mexican Government's responses to my revelations. To those responses I have had to react with the result that there has been an ever increasing cycle of accusation and counter accusation.

Throughout this sorry affair my single and over riding concern has been to secure an unambiguous and unequivocal apology, which would clear my name. I now recognise that the wording and format of such an apology, were it to be given, will have to be such as to involve no embarrassment to the Mexican government. I am mindful of that and am certain that such a wording can be found, if the will to find it is there. I will also state that I have never asked for financial compensation for the enormous damage that has been inflicted on me. Nor am I asking for it now. However, should you feel inclined to order the government to compensate me, in a properly documented manner, that would be a gesture which would be gratefully acknowledged and reciprocated in the best way possible. In the meantime, I will be settling the specifically IBM aspect of this case, but this in no way will stop my campaign to obtain redress from the Mexican government. IBM is a different matter altogether and any attempt to silence my voice in this separate matter will not be acceptable to me.

I began this petition with a plea for the intervention of your Excellency as the Constitutional Head of the United Mexican States in solving my case. In concrete terms, my plea

374

and request is that you instruct the government to order their lawyers to either contact me directly, or my Washington lawyers with a view to settling all outstanding matters between us as expeditiously as possible, before I lose control of this struggle. To demonstrate my goodwill, I will suspend my campaign ***from the moment*** I hand over this letter to your Embassy for onward transmission and shall await your decision.

I hope and plead for your intervention. I pray.

Excellency, I have the honour to remain at your service.

Kaveh Moussavi
Oxford
25 June 1993

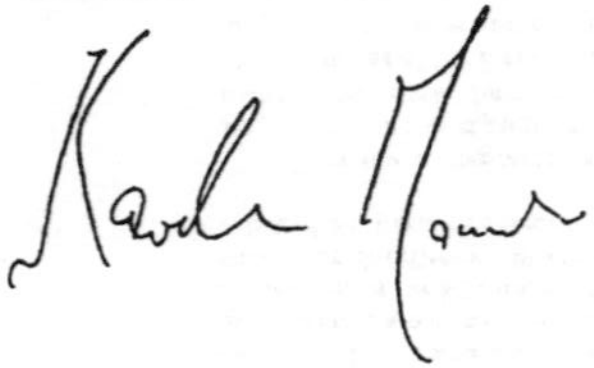

Figure 5. (Cont'd.)

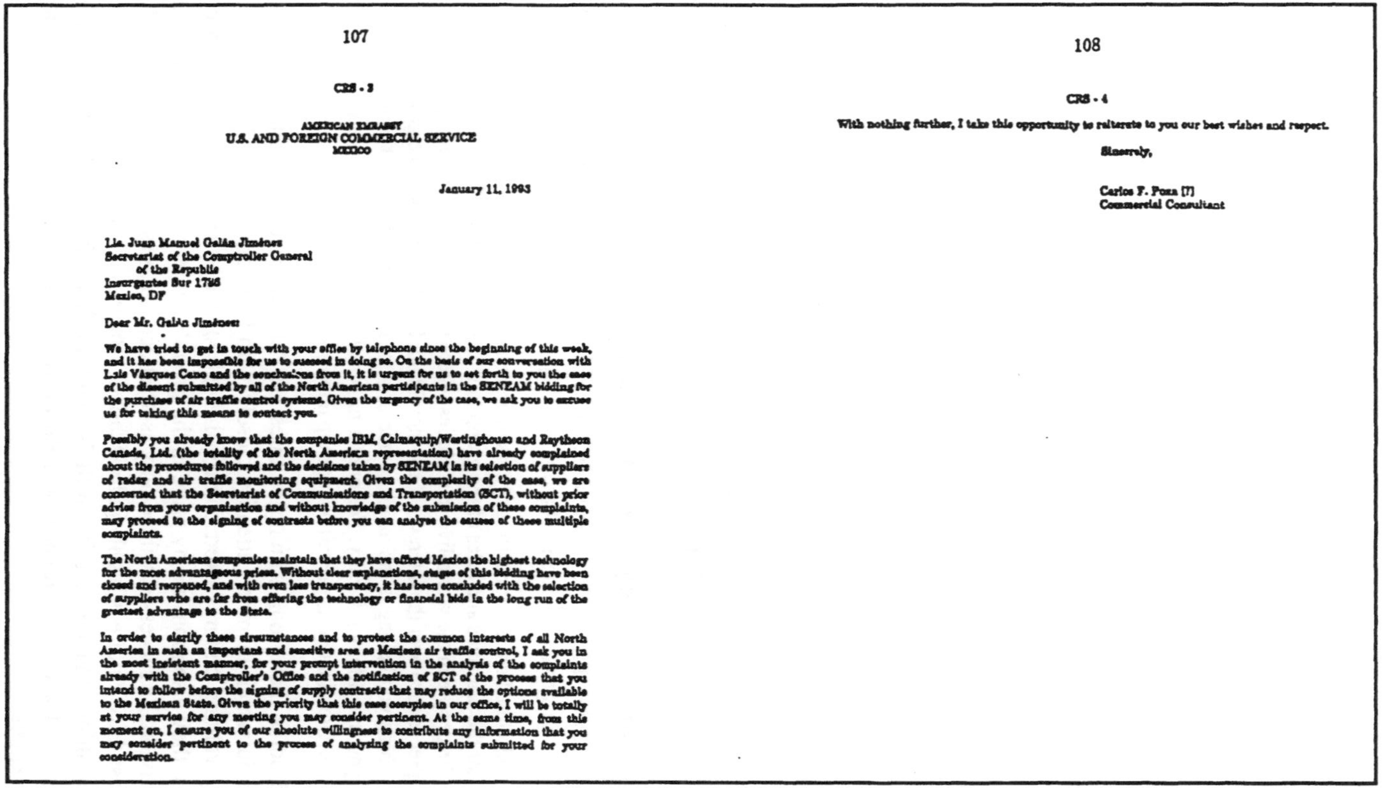

107

CRS - 3

AMERICAN EMBASSY
U.S. AND FOREIGN COMMERCIAL SERVICE
MEXICO

January 11, 1993

Lic. Juan Manuel Galán Jiménez
Secretariat of the Comptroller General
of the Republic
Insurgentes Sur 1735
Mexico, DF

Dear Mr. Galán Jiménez:

We have tried to get in touch with your office by telephone since the beginning of this week, and it has been impossible for us to succeed in doing so. On the basis of our conversation with Luis Vásquez Cano and the conclusions from it, it is urgent for us to set forth to you the case of the dissent submitted by all of the North American participants in the SENEAM bidding for the purchase of air traffic control systems. Given the urgency of the case, we ask you to excuse us for taking this means to contact you.

Possibly you already know that the companies IBM, Calmaquip/Westinghouse and Raytheon Canada, Ltd. (the totality of the North American representation) have already complained about the procedures followed and the decisions taken by SENEAM in its selection of suppliers of radar and air traffic monitoring equipment. Given the complexity of the case, we are concerned that the Secretariat of Communications and Transportation (SCT), without prior advice from your organization and without knowledge of the submission of these complaints, may proceed to the signing of contracts before you can analyze the causes of these multiple complaints.

The North American companies maintain that they have offered Mexico the highest technology for the most advantageous prices. Without clear explanations, stages of this bidding have been closed and reopened, and with even less transparency, it has been concluded with the selection of suppliers who are far from offering the technology or financial bids in the long run of the greatest advantage to the State.

In order to clarify these circumstances and to protect the common interests of all North America in such an important and sensitive area as Mexican air traffic control, I ask you in the most insistent manner, for your prompt intervention in the analysis of the complaints already with the Comptroller's Office and the notification of SCT of the process that you intend to follow before the signing of supply contracts that may reduce the options available to the Mexican State. Given the priority that this case occupies in our office, I will be totally at your service for any meeting you may consider pertinent. At the same time, from this moment on, I ensure you of our absolute willingness to contribute any information that you may consider pertinent to the process of analyzing the complaints submitted for your consideration.

108

CRS - 4

With nothing further, I take this opportunity to reiterate to you our best wishes and respect.

Sincerely,

Carlos F. Poza [?]
Commercial Consultant

Figure 6.

Figure 7 [47, p. 104]. This letter, written on February 18, 1993, by Mr. Moussavi's attorney in Washington, D.C., follows the style of a U.S. letter. The writer apparently wishes the reader to know that the legal action is being conducted on U.S. terms and not on Mexican terms. The style is direct, and concise—purely "U.S. business."

These documents reflect differences between the culture of the United States and other countries that can be described by the six dimensions that consolidate anthropological research categories of Hofstede, Trompenaars, and Hall and Hall.

A Routine Mexican-U.S. Business Situation

The final example, a business letter written by an international program director at the University of LaSalle in Mexico, to Gabriel Carranza, Director of International Projects for Mexico and Latin America at Texas A&M University, uses standard U.S. semi-block letter format. However, the writer's method of presenting his method differs vastly from that used by U.S. businesses and advocated by U.S. business communication texts. For example, the majority of the letter dwells not on the business issue but on personal interests of the writer and his relationship with Dr. Carranza (see Figure 8).

APPLYING CULTURAL DIMENSIONS: DESCRIPTION AND ANALYSIS

Dimension 1: Valuing Either Individuals or Groups

Hofstede differentiates cultures according to whether they value individualism or group-oriented behavior [42-44]. As Hoecklin explains,

> Individualism is a concern for yourself as an individual as opposed to concern for the priorities and rules of the group to which you belong. The majority of the people in the world live in societies where the interests of the group take precedence over the interests of the individual. In these societies, the group to which you belong is the major source of your identity and the unit to which you owe lifelong loyalty. For only a minority of the world's population do individual interests prevail over group interests [48, p. 35].

Mead notes that group-valuing cultures are "characterized by tight social networks in which members identify closely with their organization" [49, p. 20]. Group-valuing cultures exert control through shame at stepping outside group norms; they also motivate by pride in the group's achievement. In these cultures an individual's success is valued only as it reflects on the success of the group. In contrast, U.S. culture is highly individualistic. Business is separate from personal relationships, and one's self-worth is determined by how well one achieves professional and educational goals. Rewards in business emphasize individual achievement.

104

WILKES, ARTIS, HEDRICK & LANE
CHARTERED
ATTORNEYS AT LAW
1666 K STREET, N. W.
SUITE 1100
WASHINGTON, D. C. 20006-2866
(202) 457-7800

February 18, 1993

Lic. Manuel Galan Jimenez
El Director General
Secretaria de la Contraloria General
de la Federacion
Subsecretaria "A"
Direccion General de Responsabilidades y
Situacion Patrimonial Of. No. 220/00124
Mexico, D.F.

This is in response to your letter of February 12, 1993 in which you asked my client, Kaveh Moussavi, to provide the names of the persons who requested a payment of money and to state whether they were government employees. This letter also supplements my earlier letter of February 12, 1993.

As I advised in my letter, Mr. Moussavi is willing to cooperate with the Mexican government in its investigation of this matter. However, it is difficult for us to cooperate in an investigation where the investigating government seems predisposed not to accept the allegations of corruption. If the Mexican government wishes to have Mr. Moussavi's continued cooperation, we must be assured of the government's good will toward him and of its willingness to conduct the fair investigation that his allegations deserve.

In response to your questions of February 12, 1993, Mr. Moussavi does not know the names of the three men in question because they did not identify themselves. Mr. Moussavi assumes, but he does not know with certainty, that these men are Mexican government employees. Certainly, there are a number of circumstances indicating that they were government employees:

- They were most knowledgeable of the details of the tender;

- They said that making the requested payment would make it more certain that IBM would win the award of the contract, and they indicated an ability to influence the contract award;

Figure 7.

105

..s. Artis, Hedrick & Lane
Chartered

Lic. Manuel Galan Jimenez
February 18, 1993
Page 2

- When Mr. Moussavi refused to make the requested $1 million payment, they suggested a contribution to the government's anti-poverty program;

- In evaluating the legality of the requested payment, Mr. Moussavi asked them to demonstrate they were not government officials, but they were not able to do that; and

- During the hour-long discussion these men gave every impression of being government officials. They showed no fear of being arrested. They were willing to conduct the meeting openly in the lobby of the Nikko hotel. When Mr. Moussavi excused himself to make a telephone call to his principal they made no effort to leave, but calmly waited for him to return.

I look forward to your response at the earliest possible date.

Sincerely,

Robert X. Perry, Jr.

RXPJr:sgf

Figure 7. (Cont'd.)

UNIVERSITY LA SALLE

23 October 1995

My esteemed Gabriel:

I have wanted to force myself to write these lines since I arrived in Mexico this past Wednesday, but, as you can imagine, we were overcome in the DF and fell to the avalanche of unfinished work that we are unable to finish. But today of all days I propose to write some lines to tell you that I do not have words enough to describe how much I enjoyed all the courtesies you showed toward us. To be truthful, Gabriel, you were brilliant! I was enchanted with the A&M campus that you have told us about. . . and with just reason, because it is a marvelous place. Definitely, since I now have a son growing up, I will begin to prepare him to become a cadet when he comes to A&M. I do not know the manner in which the ideals and values become so fine in North American. Thank you for introducing us to the institution. We will never forget the visit there, which was enhanced by your great enthusiasm and incomparable energy.

After meeting with your colleagues and hearing about your grand projects for distance education, as I told you it was a coincidence that I visited an intimate friend of my daughter—Srs. Rosalia Ferreira—last Tuesday to ask me about possible work. The result was that she arranged a time with UNAM and she has strengthened her background through education by distance. Although I did not understand completely , the language of this specialty, upon hearing about this woman, I did not know what to think (perhaps by intuition. . . but for that language I am good enough) that it will be best to interview you on your next trip to Mexico see the possibility of integrating your equipment in Mexico. It seems that you have interests in the area of preparing teachiers for distance education. As it is enough to stir and to move and also as an actress of experimental theater here in Mexico (this you meant theatrical and not commercial), you undoubtedly have a good capacity for communicating and teaching. In short. . . I don't know if this interests you. I am going to ask you to revise your CV. Who can hinder us when our mutual interests coincide.

One more thing,. . . thank you for the splendid visit with your organization, Gabriel. And, please let Johanna and Cecilia know that I appreciate your friendship;. I will write them a few lines when I have one small moment. But do not allow me not to do my part, please. . . And to your beautiful wife—that it is clear why they gave her the name of "Luce"--tell her that we are looking forward to knowing her because we know that she is a great women for a great man. We all very much want your impressions of her. Happiness to all because we have mutual interests.

Hugs to all.

Signature

Figure 8.

Group-valuing organizations tend to use oral communication rather than written communication within the organization; however, individualistic organizations value written communications for their documentation value and because they assign individual responsibility.

Implications for Document Design

When writing to a business in a group culture, emphasize the relationship you are attempting to establish with the organization; de-emphasize what you want, your business objective. Instead, emphasize how your organization's liaison with their organization can enhance the business relationship between the two companies. Do not single out individuals for commendation, and emphasize "we" rather than "I." Note that in a group-oriented culture, correspondence may not include a greeting to a specific individual (not "Dear Mr. Mohav" but the anonymous "Dear Sir") because the communication is perceived to be between companies not between persons. Similarly, the letter may close with the sending company's name followed by an individual's signature "for the company" (see the Northern Traders example in Figure 1). Take a patient approach to business issues, even though these may be pressing from your perspective.

Style in documents written to members of group cultures needs to be more conversational. The sound of the message, rather than the visual clarity of the main issues, should be stressed. Note, in Figure 8, the familiar, conversational quality of the letter. It is almost as if business were the secondary goal of the letter! Courteous phrases, though formulaic, may nonetheless sound socially appropriate. In the layout of a page designed for a group-valuing audience, the identity of the organization should be primary. Photos of groups and corporate landmarks will be more acceptable than photos of individuals (except for the president and chairman). Some document designs will not work for both types of cultures.

Figure 3, Mr. Moussavi's letter to the Secretary of the Ministry of Transport in Mexico, does not discuss the problem solely in terms of Mr. Moussavi's situation. He states what has happened (Paragraph 2), but Paragraph 3 states a broader issue—that the process for accepting bids (the SENSEAM tender). Thus, Mr. Moussavi's problem has implications for Mexico that are broader than what he has experienced. Figure 5, to President Salinas, also captures the spirit of a broader concern—implications of corruption in the Mexican government, as does Figure 6, where the legitimacy of the entire bid process is questioned. As an interesting contrast in style and presentation, excerpts from Mr. Moussavi's testimony before the Congressional sub-committee are worth reading (see Appendix A). Contrasting his testimony with his letters clearly shows his ability to adopt the rhetorical stance he deems appropriate for the specific audience of each letter. We believe that this ability to determine and then adopt the needed rhetorical stance is THE most important communication competency that a student can learn.

Dimension 2: Separation of Business and Private Relationships

Trompenaar's study [45] differentiated cultures according to the extent people separate their business lives from their personal lives [see also 48, p. 44]. Some cultures separate non-business activities from business objectives, but other cultures do not make such a separation. A culture that does not differentiate among relationships bases its notion of efficiency not on time but on how well one understands others. In the United States, many people wish to separate their business lives from their personal lives. In contrasting cultures, individuals may merge business and social relationships, making individuals actually more accessible as people during business dealings. As Hoecklin observes:

> Your business partner may wish to know where you went to school, who your friends are, what you think of life, politics, art, literature, and music. This is not a waste of time because such preferences reveal character and form friendships. They also make deception near to impossible. The initial investment in building relationships is as important, if not more so, than the deal in some cultures [48, p. 45].

Implications for Document Design

Cultures that do not separate business and private relationships will expect formality, consistency in style and address, and attempts to build personal relationships in the business setting. The tone for audiences in cultures that do not separate business and pleasure should be reserved but positive, with the intention of building a suitable ethos for the writer. The U.S. writer should remember that messages can simultaneously be formal, warm, and positive. Causal style and forms of address should be avoided. Indirect messages, as opposed to direct, to-the-point-messages, are appropriate.

When designing the master pages for an audience that does not separate personal and professional relationships, emphasize formal elements, such as borders, elegant serif fonts, embossing, crests, high quality paper, and centered headings. Among the appropriate choices are coordinated designs for brochures and reports that suggest belonging to a group. Use correct titles for individuals and use these consistently.

Figure 8 provides a good example of this characteristic. This version included as Figure 8 is an almost literal English translation of the letter. Dr. Carranza knows the writer because he has dealt with him in establishing programs linking Texas A&M and various Mexican universities. Notice, however, that the writer has a business purpose—stated almost fleetingly in paragraph 2—to link the two universities via distance education. However, the majority of the letter—paragraphs 1 and 3—focus on personal items. Clearly, in Figure 8, the writer wants to strengthen the links between LaSalle University and Texas A&M, but he expects these links to occur through a happy working relationship with Dr. Carranza.

In addition, content and presentation should be conciliatory. Harmony should be maintained and confrontation avoided. Frequently, in cultures in which people's work lives and personal lives are intertwined, relationships are more important than business. Long-term organizational agendas should receive more emphasis than immediate business decisions. This application of Dimension 2 can be found in all the Moussavi letters. For example, in addressing all of the Mexican officials—Figures 3, 5, and 6—Moussavi states his problem, but he is respectful. He focuses on the moral implications of the treatment he has received.

Dimension 3: Power Distance between Social Ranks

Hofstede treats power distance as the degree of closeness that exists among members of organizational hierarchies—whether superiors consult subordinates about decisions, whether employees feel comfortable in disagreeing or questioning superiors' decisions, and whether and to what extent interdependence or authority is evident in supervisor-subordinate relationships. Thus, he says power distance is "the extent to which the less powerful members of institutions and organizations within a country expect and accept that power is distributed unequally" [43, p. 28]. In high power-distance cultures, employees manage their work according to the manager's specifications, and authoritarian attitudes are readily accepted. In low power-distance cultures, the individual is freer to follow his or her own preferences and criticize management.

United States business organizations have affinities with both low and high power distance cultures. In the United States, distance may vary significantly depending on the size and purpose of the organization. This range of power distance explains why a rhetorical approach to business communication emphasizes the importance of determining the relationship between writer and reader(s), the job-related distance between writer and reader, the appropriateness of the communication mode selected, as well as the appropriateness of the message itself.

Implications for Document Design

In high power-distance cultures, using correct forms of address can make a difference: knowing specifically to whom to address the report or letter, the title or rank of that person, which names to place on the distribution list, and what rank each decision-making individual holds. Establishing the correct tone in addressing the intended reader(s) and thus recognizing the appropriate distance needed between writer and reader(s) is also important. Therefore, tone in high power-distance cultures may need to be more formal if the writer is preparing a document for an audience who holds a relatively superior position. Official formats, consistent graphic hierarchies, and communications designed for special occasions, such as commemorative scrolls, announcements, and commendations, are appreciated. In contrast, in low power-distance cultures, strict recognition of business hierarchies and formal address gain less favor. A wide variety of formats may be used, and informal layouts and casual typefaces will be acceptable. The style of the message can also be more casual.

The Moussavi letters, Figures 3 through 7, illustrate this dimension in a spectacular way. In addressing the Mexican officials, Mr. Moussavi is appropriately respectful (perhaps a bit obsequious, if we contrast his correspondence with his testimony). He recognizes their superior rank. Notice his forms of address in Figures 3 and 5. Note that he is oblique, eloquent (pompous from a U.S. perspective) in presenting his petition. He knows that high-ranking Mexican officials will expect this approach. Compare his approach with Carlos Poza's letter (Figure 6) to the Secretary of the Comptroller General's office. This person has a much lower rank, and the letter reflects that fact. The message is still indirect, but the message lacks the prolixity of Figures 3 and 5, but is still appropriately formal.

Figure 4 contrasts starkly with Figures 3, 5, and 6. In his letter to his Mexican Consulate General in London, where Moussavi lived at the time, Moussavi does not mince words. He is blunt, terse, and specific in his demands because he now believes that he is dealing with a Mexican representative on non-Mexican turf. Moussavi is almost threatening in his demands. Rather than elegant variation (as in Figures 3 and 5), we have direct, staccato demands.

Dimension 4: Universal or Relative (Particular) View of Truth

Universalism, according to Trompenaars [45], applies where people believe that what is true and good can be discovered, defined, and applied everywhere. In contrast, particularism values circumstances and relationships over abstract rules. Universalism applies to cultures where societal values are more important than personal obligations. Written documents such as contracts are highly valued, as these documents contain the truth of the relationship.

Implications for Document Design

For the universalist culture, be as specific and concrete as possible. Clarity and precision via format, diction, syntax, and usage will be valued. In argument structure, reference to agreed on principles leads to interpretation of specific details; alternative possibilities need not be given as much attention. Drawings of normative cultural types, especially time-honored ones (a mother feeding her child, an executive making a gesture of instruction, a person making a gesture of respect) may be successfully associated with products or services rather than photos of specific individuals. Traditional layouts that may look busy to Westerners will be acceptable. (See the Gannon Dunkerly example in Figure 1 and Northern Traders example in Figure 2.)

Figures 3 through 7 also allow examination of differences in universality versus specific culture. Mr. Moussavi believes, from his U.S. perspective, that requiring a bribe from a company that wishes to submit a proposal is wrong and clouds the fairness of the bid process. He is outraged by the threats on his life and the lives of his family because he reported the bribe. However, he also knows that the Mexican government will likely take a different attitude toward the entire affair. While his attorney deals with the issue in a U.S. way—focusing on the legal issues of the attacks on him by the Mexican press—Mr. Moussavi appeals to the Mexican oligarchy's own sense of honor, their standards of what is reprehensible. To Gamboa Patron, Moussavi describes the "libellous [sic] propaganda drive" unleashed against him by the Mexican press. He states (paragraph 3) that he knows that Patron will want to investigate the bid process. In his request to President Salinas, he states that he asks only for an apology to clear his name. He does not ask that the "mordita" system be changed. In short, in addressing different readers, Moussavi uses different rhetorical appeals. What is "true" for one situation, what argument will be appropriate, will not apply for another.

Dimension 5: Whether the Entire Message is Contained in the Text

Hofstede's widespread influence on international research has led other international business scholars to extend his value system. O'Hara-Devereaux and Johansen, echoing research by Hall and Hall, suggest five lenses for examining the culture in a society: language, context, time, equality/power, and information flow. They suggest that these five lenses form a holographic, or dynamic interrelationship [50, pp. 50-51]. Context, however, seems the most viable lens for exploring a culture's response to communication patterns. Some cultures express in words or images all of the relevant details needed to make sense of an explicit message: The document must contain the full meaning. In other cultures, the meaning of the verbal message is not limited to the text itself: other issues, events, and personal relationships must be considered for the text to be understood.

Implications for Document Design

If prospective readers will typically look beyond the text, the writer can include minimal explanation. However, the discussion may still be "wordy" by U.S. standards of stylistic economy. According to O'Hara-Devereaux and Johansen, "Since words have relatively less value, they are spent in great sums" [50, p. 56]. However, if readers expect the text to contain everything needed to make sense of the message, the writer should express situational details fully and use language with precision and economy. Every word is meaningful. The text should be able to stand alone apart from the context in which it was created.

In Figure 8, the writer expects that Gabriel will understand what type of distance education linkage he is suggesting. In this situation, the letter opens the way for future face-to-face discussion. In contrast, in Figure 2, the writer is explicit about what information Northern Traders is seeking.

Dimension 6: Whether Uncertainty is to be Avoided or Accepted

Hofstede's uncertainty avoidance index measures how much uncertainty a culture tolerates about the future [43, pp. 113-114]. Members of cultures that avoid uncertainty appear anxiety-prone and feel the uncertainties inherent in life as continuous threats that must be fought. In the organization, they fear failure, take fewer risks, resist change, and place a premium on job security, career patterning, and company benefits. The manager is expected to issue clear instructions, and subordinates' initiatives are tightly controlled. Employees in cultures that dislike uncertainty accept formal procedures, wide power distances within hierarchies, and highly structured organizations [49, pp. 18-19]. In contrast, members of cultures that accept uncertainty are more likely to take each day as it comes. Conflict and competitiveness can be used constructively, and dissent will be tolerated. Needs for written rules and regulations are relatively few, and if rules cannot be kept, they can be easily changed.

Strong value systems, Peters and Waterman noted in *In Search of Excellence*, negated the need for certainty in successful companies. Without exception the dominance and coherence of culture proved to be an essential quality of the excellent companies. Moreover, the stronger the culture and the more it was directed toward the marketplace, the less need there was for policy manuals, organization charts, or detailed procedures and rules. In these companies, people way down the line know what they are supposed to do in most situations because the handful of guiding values is crystal clear [51, p. 179].

Implications for Document Design

In cultures that accept uncertainty, written documents may be less problematic than they are in cultures that avoid uncertainty, where documents are valued for documentation and governance purposes. Forms, tables, and many appendices serve the needs of high-certainty seeking cultures. Images associated with certainty, such as flowcharts that make outcomes clear, elaborate borders (the kind on stock certificates), high quality paper, traditional layouts, precise physical images, such as photos, boxed quotations or principles, and presentation techniques that look permanent (embossing, engraving, framing) will be valued. Figure 4, written to an English reader, is much more specific than the letters to the Mexican hierarchy (Figures 3 and 5). The U.S. culture, a low-context culture, seeks certainty and uses documents to eliminate uncertainty, where Mexico, high-context culture, is more accepting of uncertainty. None of the letters that Moussavi submitted to support his testimony are written as documentation. Instead, they are appeals to clear his name, not in terms of his well-being, but in terms of the Mexican government doing what is honorable for individuals in their position.

THE USEFULNESS AND MUTABILITY OF VALUE DIMENSIONS

Value dimensions, we want to reemphasize, are useful in helping writers understand how to begin to design documents for a specific culture, but they are always subject to change. Clearly, these value dimensions are not independent of other factors. As more companies become multinational, and as U.S.-based companies launch and expand international operations, a company with confidence in its national origins and corporate culture will likely dominate the culture of the company's operations in other countries. Thus, technology selected and implemented by the parent company, the presence of an entrenched communication style in the parent company's corporate culture, the ability of employees to use technology, and the flexibility of the technology to adapt to cultural preferences of users, each of these parameters will shape the written communications of a company.

However, we also note that a number of the value dimensions seem to produce common document design characteristics. For example, many cultures prefer aesthetic and fluid, rather than concise, direct style. Many cultures are less direct in stating the purpose of their message, while they are also less exacting in expecting precision in stating "all the facts." These communications are also conciliatory and avoid confrontation. This observation does not mean that all messages in any culture will be oblique and taciturn. The value dimensions simply alert a writer that research has strongly indicated these values are pervasive in some cultures.

In addition, the importance of business objectives, in contrast to relationships among those involved in business transactions, may vary among organizations; however, in countries like Mexico and Japan, the importance of human relationships cannot be ignored. We also believe that changing economic realities in countries like Mexico and Japan may have a bearing on the relevance of these dimensions. Given the rapid economic chances within the global village, the value dimensions themselves are likely undergoing transformation. When Honda experienced continuing loss of market share, changes consistent with U.S.-style management that would have been unthinkable during a more successful era, were adopted by the company's new president [52, pp. 92-100].

Thus, the six value dimensions for approaching and understanding other cultures should be scrutinized both for their ongoing influence and for their interaction with other influences. We believe that economic reality powerfully impacts communication. Furthermore, electronic communication (e-mail) will likely transform the design of international messages in a significant way. The role that technology plays in shaping any culture's communication has yet to be ascertained. Students can use the six value dimensions along with knowledge of a country's economy and technology infrastructure to modify communication strategies for each situation.

A COMPARATIVE GUIDE TO INTERNATIONAL DOCUMENT DESIGN

The problem in teaching international business communications lies in the cultural issues faced by U.S. students. Most are so accustomed to the U.S. culture that they are unaware of its characteristics. We maintain that students need to recognize the characteristics of their own culture, in terms of the dimensions discussed, before they can work with the dimensions as they apply to other cultures. Appendix B is a guide to help students see comparisons between the U.S. culture and non-U.S. cultures and to show how U.S. documents may differ from those in other cultures. Admittedly, the guide is a generalization, but it shows how U.S. document design, the focus of the standard business communication course, contrasts with document design found in the documents of many non-U.S. countries. The letters (Figures 1 through 8 just discussed) can be used to help students understand the design concepts.

In planning and designing business communications, U.S. professors have traditionally taught students to determine, as accurately as possible, the reader(s) they will be attempting to reach, the profile of the reader(s), and the goal of the message. From this analysis, writers select and shape content, structure, and style of the message. In designing documents for international audiences, we believe students should be taught the same strategy, but it should be broadened to include the considerations we have presented here. First, students should be introduced to value systems on which the guide is based. Instructors should have available the diagrams that Hofstede generates in this studies as well as Trompenaars' discussion and visual diagrams in *Riding the Waves of Culture* [45]. A number of books, such as *NAFTA: Managing the Cultural Differences* by Moran and Abbott [53], incorporate Hofstede's values dimensions and diagrams into their discussion of the culture. Many of these kinds of books provide case scenarios that can be extended to require letter and report exercises. Books like Richard Mead's *Cross-Cultural Management Communication* [49] use Hofstede's theory as a basis while providing writing and speaking exercises.

But whatever sources may be used, students should first understand how the U.S. culture, their native culture, fits within the paradigm. We believe that most U.S. students do not understand the characteristics of their own culture and the ways that the six key dimensions are reflected in U.S. culture. Students often fail to understand how the guidelines for letters, reports, and page design, which they are learning, are applicable only for U.S. documents. Thus, students need to recognize the characteristics of their own culture and understand that the way U.S. documents are written often differs from the way that documents are written for individuals in other cultures. The guidelines, included as Appendix B, provide a supplement to help students see how their culture may differ from other cultures in the design of written communication. Once these differences are discussed, students can then begin to answer the following analytic questions before responding to an international business writing assignment.

Pre-Writing Questions

- Who is/are the reader(s) of this message?
- What do you know about them? age? interests? education? job responsibilities? title?
- What are their particular characteristics, as gleaned from messages they have written or encounters you have had with them?
- If the document is being directed to a reader in a non-U.S. culture, what are the characteristics of this culture along the critical six dimensions? individual or group-oriented? separation of business and private relationships? high or low power distances? see truth as universal or as relative to particular circumstances? entire message usually contained in the text? seek certainty or accept uncertainty?
- How fluent are your readers in written English?
- What is the situation that has led to your need to write this document?
- What purpose do you hope to achieve? What do you want to happen as a result of this document?
- Based on the broad value characteristics of the culture, what choices do you need to make about structure—deductive or inductive? organization of ideas? degree of specificity about business purpose? type of information about you and your organization? quantity of detail about you and your organization? stylistic features such as sentence length, word choice, address protocols, tone, formatting techniques, and graphics?

BUSINESS COMMUNICATION AS CULTURAL EDUCATION

Cultural dimension approaches, such as Hofstede's, Trompenaars', and Hall and Hall's, are obviously limited in their ability to provide specific communication instruction for specific cultures. However, the strength of this cultural approach lies in its potential for helping U.S. students initially understand the characteristics of their own culture and some of the basic ways in which Western cultures differ from non-Western cultures. These differences, we believe, despite additional characteristics of any culture's written communication, determine the broad differences in document design. These differences can also guide pedagogy for teaching students to design business communications for these cultures. Thus, in broadening business communication instruction to include international communication design via cultural value dimensions, we are showing students that:

1. Ideology underpins the discourse system of the culture.
2. Form and style of discourse arise from ideology.
3. The relationships we establish via communications are a product of ideology.

As Peter Drucker wrote nearly three decades ago, "Information is not communication. It is perception" [54]. When we encounter other social systems, they have already given names to themselves, decided how they want to live, and how the world is to be interpreted. We may label them if we wish, but we cannot expect them to understand or accept our definitions, unless these correspond to their own. We cannot strip people of their common sense constructs or routine ways of seeing. They come to us as a whole system of patterned meanings and understandings. We can only try to understand, and to do so means starting with the way they think and building from there [Schutz, 1970; cited in 48, p. 9].

The new world economy means that students must broaden their understanding of non-U.S. cultures. Students need help in comprehending that people in all business cultures do not react to the same contextual cues students know. Readers from other cultures actively select, interpret, and create their environments based on what their cultures enable them to select, interpret, and create. The conclusions they reach are based on their perceptions of the evidence supplied by the physical, social, and moral universe in which they live. Because business communication provides a realistic, pragmatic environment for studying major characteristics of both international and national cultures, business communication can be viewed as a new, integrating element in liberal arts education.

APPENDIX A

STATEMENT OF KAVEH MOUSSAVI, IBM'S FORMER POLITICAL AGENT IN MEXICO

Mr. MOUSSAVI. You did, sir, yes.

Mr. Chairman, sir, ladies and gentlemen, I note that in the press release it is actually mentioned that I am here under subpoena, so I don't have a prepared text. I am going to speak to you for 10, 15 minutes, on my experiences in Mexico.

Since appearing before another august committee of this House, Mr. LaFalce's Small Business Committee, Mr. Chairman, you might be interested to know that I have now been promoted to the dubious status of public enemy number one in Mexico. I have the documents here. I shall very much like to put it into the record.

My sole crime, of course, is that I dared to expose an act of corruption. Rather naively I believed in the glossy brochures put out by the Salinas administration and their cronies and representatives in Britain, Western Europe, and the United States. I am on record as being an extreme Salinista. I have nothing against free trade or indeed against NAFTA as such.

I was one of the people who actively participated in promoting the image of Mexico under Carlos Salinas de Gortari in promoting seminars and so on and so forth. Alas for illusions, Mr. Chairman, the reality, when you confront it, is very, very different.

I would ask that the members of this committee rather than simply reading what appears in the press try to do a little more of what is being done today, confronting a few businessmen who have had firsthand experience of Mexico. The campaign of defamation, character assassination, and more than that, and I will come to that in a moment, which I have had to suffer over the last 9 months is well-documented.

Truly, I can say while the Salinas administration has now promoted me to the status of public enemy number one in Mexico, they themselves have sunk to newer depths of depravity. They have truly set new standards in gutter journalism. I will submit into the record of this hearing later a number of the articles that have appeared about me in Mexico, and I leave it to your own judgment.

On a happier note, after my appearance before the previous committee in which I put into the record a document which catalogued the steps which Her Majesty's government in Britain and the British police have had to take to protect my life and that of my family against the threats which I have absolutely no doubt come from one direction, and I need not spell it out. Surprisingly, after I said that if anything untoward were to happen to a member of my family or myself I would hold Carlos Salinas de Gortari personally responsible, I am happy to announce that since my appearance before that committee all intimidation has ceased.

I leave it to you, Mr. Chairman, to draw your own conclusions. Let me reiterate I have never been a part of an anti-NAFTA, anti-Salinas, anti-Mexico campaign, and anything that I say here I hope, I hope, and I note that there are Mexican journalists here, I earnestly hope would not be interpreted as an anti-Mexican crusade. I am not anti-Mexican, I am not anti-NAFTA as such, though I certainly have doubts about the credibility and the veracity of the undertakings that are given by a government that shows absolutely no respect for the rule of law in its own country.

It shows no inclination whatsoever to abide by its own laws and its own constitutional requirements and so on. I have serious doubts about entering into arrangements with a government that is of that nature. I think it will be best to call a spade a spade and call the Salinas administration really what it is. It is an undisputed fact, Mr. Chairman, that is not even denied by the most vociferous advocates of his regime that Carlos Salinas de Gortari came to power through electoral fraud, electoral fraud on a massive scale.

It is not something that I say, it is something that is agreed to even by the respected conservative London journal, *The Economist,* which has been a consistent supporter of the Free Trade Program and the Salinas administration from day one. In the survey, a detailed survey that they carried out of Mexico, they readily concede in the same breath that they called Carlos Salinas one of the greatest statesmen of the 20th century, they admitted that he came to power through electoral fraud.

The baptism of that government was through fire, indeed. Something like 200 supporters of the opposition have been murdered in the last 4 or 5 years, and coming up to this current election God alone knows what is going to happen there. My testimony is based on my own direct experience of a number of cases, but in particular the case that in Mexico has come to be known as the IBM Moussavi-Caso Lombardo scandal.

The facts of the case, I think, are quite well known. I will run you through them very, very briefly. I was appointed IBM's agent to assist them with their effort to win a contract for air traffic control in Mexico. IBM has an office there, employs, I believe, 2,000 people or so, they have been there since the early part of the century, and yet for reasons that I have not been able to totally fathom, but I have my suspicions, IBM corporation's office in Mexico did not wish to participate in this tender.

The reason for that, it is my strongest suspicion, was that they knew that the tender is not going to be clean. It is not going to be a fair tender. It is going to be corrupt. My experience later on in the events that I will refer to in a moment that took place on the morning of November 9 at the Nikko Hotel confirmed somewhat dramatically that public tenders in Mexico are allocated, shall we say, not on the basis of merit, but on the basis of who are the traditional winners, and I will come to that in a moment.

We agreed to participate in this tender. We put a bid at the right time. We were advised that—we were unofficially advised that—we were doing extremely well. By the middle of the tender, and I would like to put on record IBM and I had absolutely no doubt and certain statements have been made in Mexico since my last appearance suggesting that IBM would like to disassociate themselves from this statement. I certainly would challenge them to deny that there is ample documentary evidence which would seem to indicate that at least it was their opinion at the time that the tender was being run corruptly. That is to say in the identification of corrupt individuals in the organization that was running the tender.

If you wish to have details of that, I should be more than happy to oblige later. The tender itself was for an air traffic control program, modernization of air traffic control. We participated in it in the normal course of events. I will not go into the details as such, only to say that halfway through the tender I was the recipient of a number of solicitations for assistance, shall we say, assistance to help to win the contract.

I did faithfully and dutifully report these matters to IBM without ever going into detail. I simply said that this is what the situation is. People are asking to give us assistance. IBM Corp., agreed, and they did authorize me to investigate this matter further. I note that in Mexico certain statements have been attributed to IBM which I have absolutely no doubt could not have been made by IBM Corp.

For the record, I believe the company to be an extremely honorable company. I believe the company to be staffed by extremely honorable individuals. Nevertheless, they did authorize me. I have documentary evidence in the form of a letter dated November 4 from a senior IBM officer written to me allowing me to go down to Mexico to investigate these solicitations.

For the record, and for the benefit of the Government of Mexico, I will put that letter into the record here today. I arrived in Mexico City, Mr. Chairman, on November 8, and as prearranged with my interlocutors I did, indeed, meet the relevant people who had come to see me. Their intention was absolutely clear. They made no bones about the fact that they wanted a contribution of $1 million to President Salinas' solidarity program. I did report the incident to IBM.

I specifically mentioned the name "solidarity." I specifically mentioned the request for a political contribution. I specifically mentioned to IBM that the people were in my considered opinion government officials.

In the confusion that has been created by the cacophony that comes out of Mexico through the government-controlled media, IBM's position and my position has become extremely confused. There can be absolutely no doubt and the documentary record speaks for itself that I did mention that I had been approached and requests were made for political contributions. I repeat, political contributions. I say this in the light of the later attempts by the government in Mexico to obfuscate the circumstances surrounding this tender, and they have attributed all sorts of absurd motives to me and the story surrounding this.

In the event we were unable to make a deal with these gentlemen, who were asking for these payments for the simple reason that at the end of the day I made it a condition that I had to know who they were, and if they were government officials I could not deal with them. For the record, I will say my purpose in going down to Mexico was to discover whether I could make a legal payment, that is to say whether we could come to an arrangement with expert consultants who could assist us in the course of a normal tender.

There were unable to demonstrate to me that they were not government officials, and therefore despite the best will in the world and despite our determination to win this contract we were unable to make a deal with them.

I would be very happy if specific questions are put to me on this aspect of it later. I was warned at the time that if we did not pay, the tender would be swayed away from us. I reported this faithfully to my controllers at IBM at the time.

Mr. Chairman, precisely 10 days later on the morning of November 19 the Government of Mexico announced that none of the contenders were compliant with the requirements of the tender. This was a strange suggestion. Here a Third World country was telling the companies which were on the leading edge of technology that they didn't know what they were doing, but we were rather bemused by that.

A few days later another tender was put out. The terms of the tender were so dramatically changed that in the considered opinion of IBM officers and myself, only the traditional winners could have won this tender. As it was, the case was closed. It was awarded to a nationalized company, the French company, Thomson, and another State company, Italy's Alenia, another nationalized company.

The award was made on December 28. At the time, Mr. Chairman, sir, I think it is important that I reiterate these IBM officers and myself were so outraged at the Government of Mexico that we decided that we would file a formal protest with the government, but we soon discovered that we were not the only people. I can tell you, Mr. Chairman, that in my fairly extensive experience as a businessman I have never, ever, ever participated in a tender in which every single company files a protest. Every single company protested this tender.

IBM Corp., my client, and all the other companies, Japanese companies, British company, British-German company, a Canadian company, and another American company, but to our amusement and surprise we found that the United States Embassy also protested. We found that the Japanese Embassy protested. We found that the British Embassy protested. The Canadians did even better, Mr. Chairman.

The Canadian Trade Minister himself, Mr. Michael Wilson, formally wrote to his counterpart in Mexico, Mr. Caso Lombardo. I can tell you that by itself should be enough to raise an eyebrow or two about the nature of this tender, and it wasn't an ordinary tender. We are talking about air traffic control. We are talking about the lives of ordinary people who are going to be flying over national territory of Mexico.

Needless to say, the Government of Mexico brushed aside all these protests, and IBM and I decided that this was too much. We were going to take this protest to public domain. I briefed the *Financial Times* with the help of IBM who produced a press memorandum for me and the *Financial Times* produced a report on February 3. If I were to cut the story here and now, ladies and gentlemen, and if it were just to say begin from here and watch the knee-jerk visceral gut reaction of this government toward a forthright statement of a witness who is prepared to come forward, risk everything in Mexico simply on the basis of the belief that he has belief in the goodwill of the Government of Mexico, if you simply watch the reaction of that government, I think it would help to disabuse anyone of any illusion that this is a reform government, that this is a government that is determined to uphold the rule of law, to preserve the principles of due process, to respect at least to a certain degree the requirements—the basic requirements of what goes into a judicial investigation.

If you just catalog the reaction of the Government of Mexico from the afternoon of February 4, the day after the *Financial Times'* report on my statement as well as their own investigation and analysis of the tender appeared, I don't think anyone in their right mind could conclude that this is a bunch of people one could do serious business with. The reaction was immediate; it was hostile. The campaign of defamation that began against me was just unbelievable.

On the 4th the government officially put out a statement in which the presumption that I was lying was total. On the 5th they put out yet another statement, effectively saying that the Attorney General of Mexico, through the Mexican Foreign Ministry was going to come after me. I would be extradited and all sorts of other things.

On the 6th, 3 days later, the Minister himself did even better. He appeared on television and denounced me as a liar and already he acted as the prosecution, judge, and jury at the same time. I was already sentenced to a prison term within 3 days. What possible investigation the Government of Mexico could have carried out into a forthright allegation by an agent, a businessman who had been there who knew exactly what was involved, I don't know.

Counsel tells me I am talking too much and I should cut it. The reaction was like that. I was already condemned. This was on the 6th. But to my repeated request that there should be an immediate investigation, having condemned me already, having already said I should be put in prison, the Government of Mexico, on the afternoon of the 12th, please remember the dates, 6 days after they had already condemned me, they write a half-a-page fax, half of one page to my attorney, Mr. Bob Perry, asking him, half of one page, which is the sum total of the investigation that they had carried out into my allegations, asking him to provide the names of the men who came into the Nikko Hotel to interview me, knowing full well that I didn't know the names of these people.

I was unable to get them for the simple reason that they didn't want to let it be known that they were government officials. Mr. Perry replied to them on the 18th. Immediately, you could see they had the reaction ready, on the morning of the 19th the Government of Mexico put out a statement saying there is absolutely no evidence for what I am saying, simply because I didn't know the names of the people who had come to solicit the bribe from me.

As a rule I don't know of any people who usually come and ask you for a bribe and they bring television cameras and their lawyers and commission contracts and so on. However, as far as the logic of the Government of Mexico goes, Mr. Chairman, the fact that a witness to a crime does not know the names of the criminals of necessity means that no crime has taken place.

If you, Mr. Gonzalez, sir, walk out of your office tomorrow and you witness a bank robbery in the street and you immediately phone the police and the police stop you and ask you for the name of the robbers and you don't know the names, they conclude that the bank has not been robbed. More than that, they conclude, Mr. Gonzalez, sir, that you are a liar, which is exactly what they have done with me.

Unable to seek judicial redress in Mexico, and I tried very hard, I retained an attorney, and I asked him to issue immediate defamation proceedings against the Government of Mexico and in particular against the Minister. His response was, Mr. Moussavi, your naivete is really touching. This is Mexico. This is not the United Kingdom or the United States.

I was obliged to issue a defamation suit against the Minister in London. When I did, in the English courts, knowing full well that they could not bribe the judges there or intimidate them, their reaction was instantaneous. Not only they went on television and condemned me again, but very soon after that a senior Foreign Ministry official, Mr. Eduardo Ibarrola, was sent to London, and I have the documentary evidence here, to seek a meeting with me to arrange a comprehensive accommodation. I think you have guessed it, Mr. Chairman, the idea was to bribe me into silence.

They basically offered to assist me with any other contract that I may wish to have in order to shut me up. This, of course, I rejected with the contempt that it deserved, and I told Mr. Ibarrola had this been a tender for 10,000 bars of soap for the Guadalajara municipality, I might have been tempted, but we are talking about the lives of ordinary people. We are talking about an unsafe air traffic control system. We are talking about no ordinary tender.

Needless to say, I did not get the investigation. I didn't get the apology, so I exposed the matter to the press. The result was an immediate denial by the Government of Mexico that they had tried to bribe me, but as I said in the previous committee, the Lord does work in mysterious ways. Six drafts, Mr. Gonzalez, sir, ladies and gentlemen, six drafts of the denial letter of Mr. Eduardo Ibarrola which was published in the Mexican newspapers appeared on my fax machine from a well-wisher working the Mexican Foreign Ministry who wrote to me saying, Mr. Moussavi, with profound admiration for what you are doing, with utter disgust for my corrupt government, I would like to assist you to expose these people.

When you look at those drafts I don't think anyone, any fair-minded person can fail to recognize that the government was desperate to clean up after their efforts. Basically, it would be regarded in the United States as an attempt to interfere with a key witness, to basically try and bribe me. I published those, needless to say. For that I have now become public enemy number one in Mexico.

My life has been threatened on a number of occasions; my children's lives have been threatened on a number of occasions. The last time I was going to come to appear before—after responding to the subpoena issued by Mr. LaFalce's committee on the morning of Saturday, I received a telephone call which mercifully the police eventually managed to trace in Britain in which they said if you appear before the United States Congress when you return you will be one child less, Mr. Moussavi.

I leave it to you to imagine what my state of mind was, Mr. Chairman. My motives have been questioned as to why I am here today. I am glad that you mentioned that I am here under subpoena. I am also here, Mr. Chairman, for one reason, one reason alone. The more the press and the more the American public hear about my case, I should think I am going to be a bit safer.

The conclusion, I would say, is that here we have a situation in which an air traffic control contract has been awarded, and the issue goes beyond the air traffic—this particular air traffic control contract. It wasn't very big. The issue goes to the heart of the nature of public procurement in Mexico. It goes to the heart of the nature of the judicial process in Mexico. It goes to the heart of the possibility or rather the impossibility of obtaining judicial redress in Mexico, and it should be a solitary lesson. It should be a warning to anyone, any businessman who goes into Mexico, sees corruption and dares to denounce it. This is what happens to them. Thank you very much.

APPENDIX B
Understanding Differences Between United States and Non-U.S. Written Business Communications

Written Communication Strategies

UNITED STATES BUSINESS CHARACTERISTICS	NON-WESTERN BUSINESS CHARACTERISTICS
IMPORTANCE OF INDIVIDUALS VS. GROUPS	
Insistence on written communication to show accuracy and individual responsibility for document	**Preference for face-to-face communication**
• Style in written document is structured by paragraphs and business protocol;	• Style in written documents should be conversational and reflect oral, spoken language
Values rapid, efficient pace of communication	**Values slower pace of communication**
• Conciseness valued. Eliminate irrelevant information. Eliminate wordiness, vague terms, Use C-B-S style (clarity, brevity, simplicity)	• Conciseness not valued; choose conversational, friendly (non-business) approach.
VALUE OF PRIVATE AND BUSINESS CONCERNS	
Organizational tasks/goals are highest priority. Business before relationships	**Organizational and personal relationships are more important than business. Relationships have highest priority.**
• Emphasize recommendations, procedures,	• Emphasize company background and rationale for business actions
Work and personal lives separated. Separate emphasis on each individual's responsibility in achieving business goals	**Work and personal lives intertwined. And life intertwined. Emphasis on the responsibility shared by group**
• Message is strictly business; personal life is separate from business life. Tone of document: objective, direct, and precise	• Message links many aspects of the culture personal/public/corporate lives shared. Tone of document: subjective and oblique.

APPENDIX B (Cont'd.)

UNITED STATES BUSINESS CHARACTERISTICS	NON-WESTERN BUSINESS CHARACTERISTICS
EFFECTS OF POWER DISTANCE	
Values informal communications	**Values formal communications**
• Less emphasis on titles, consistent use, and occasion for communication.	• Emphasizes use of formal titles, forum, and audiences for messages.
Careful use of distribution lists to limit access and readers. Decision making via designated leaders	**Interdependencies crucial to intra-/inter-organizational communication. Decision making via consensus**
• Hierarchies in the organization and in those receiving copies. Focus on those responsible for acting on the message.	• Focus message on all groups that are affected by the message.
Short-term organizational agendas shape messages.	**Long-term organizational agendas shape messages.**
• Emphasize action required and time tables for achieving goals.	• Discuss business in terms of larger human and company concerns.
VIEW OF TRUTH	
Values analytical, logical, denotative written communication.	**Values intuitive, aesthetic written communication reflective of oral discourse**
• Emphasize structured, logical messages; use of headings/page design.	• De-emphasize structured response and use of headings/page design.
Truth is defined by written agreements and facts that can be seen in one way. Truth is found in the document, which has one meaning. Messages are to the point; often abrasive.	**Truth may be modified by changing realities; truth not confined to written document.** Truth is a product of several perspectives; written, spoken, felt. Messages often evade issues; are closed and introverted.
• **Messages are binding**; they contain truth, which can be defined and agreed upon by all parties.	• **Messages may be formal, but they are revocable.**
• Style in written document is structured by paragraphs and business protocol; linear discourse. (Figure 2, Figure 5)	• Style in written documents should be conversational and reflect oral, narrative discourse; non-linear discourse (Figure 4).
Because business is the major priority, messages should be direct. Honesty in communication is valuable.	**Harmony should be maintained and confrontation avoided. Conciliation is valuable.**
• Content and presentation are argumentative. State the issues squarely but tactfully. Use design to highlight facts and desired outcomes.	• Content and presentation are conciliatory. Control tone to allow everyone to save face in unpleasant situations; avoid blame on individuals or groups.
Agreements legally binding.	**Written agreements not binding; meaning can change depending on changes in circumstances of the company.**
• Message should aim to capture the desired goal of the communiqué; ambiguity in goals not desirable. Conciseness more important than elegance.	• Messages should be oblique; building relationships carefully and futuristically with business in the background. Elegance more important than brevity.

APPENDIX B (Cont'd.)

UNITED STATES BUSINESS CHARACTERISTICS	NON-WESTERN BUSINESS CHARACTERISTICS
CONTEXTUAL COMPLETENESS	
Low-context communication; work and life separate; emphasis on individuals. Resonsibility belongs to specific individual(s).	**High-context communication; work and life intertwined. Emphasis on the group responsibility shared by members.**
• Message is strictly business; personal life is separate from business life. Tone of document: objective, direct, and precise.	• Message links many aspects of the culture: personal/public/corporate lives are shared. Tone of document: subjective and oblique.
UNCERTAINTY ORIENTATION	
Values direct communication	**Values indirect communication**
• Highlight main point; begin with most important ideas first. State point clearly and objectively.	• De-emphasize "main point" or place it at the end of the message. Message should be obliquely stated.
Deadlines/commitments are firm	**Time/deadlines are relative and flexible**
• Time requirements specifically given	• Time not pressed in the message
Approach is practical; action-oriented problem-solving approach	**Text is indirect and theoretical; specific implementation indirectly stated**
• Goal should be visually clear; bulleted lists	• Goal required may be hard to explain

REFERENCES

1. U. Connor, *Contrastive Rhetoric: Cross-Cultural Aspects of Second-Language Writing,* Cambridge University Press, Cambridge, England, 1996.
2. U. Connor, A Contrastive Study of Persuasive Business Correspondence: American and Japanese, in *Global Implications for Business Communications: Theory, Technology, and Practice: 1988 Proceedings of the 53rd National and 15th International Convention of the Association for Business Communication,* S. J. Bruno (ed.), School of Business and Public Administration, University of Houston-Clear Lake, Houston, Texas, pp. 57-72, 1988.
3. C. Kostelnick, From Pen to Print: The New Visual Landscape of Professional Communication, *Journal of Business and Technical Communication, 8*:1, pp. 91-117, 1993.
4. *Global Networks: Computers and International Communication,* L. M. Harasim (ed.), MIT Press, Cambridge, Massachusetts, 1993.
5. L. P. Driskill, Collaborating across National and Cultural Borders, in *International Dimensions of Technical Communication,* D. C. Andrews (ed.), Society for Technical Communication, Arlington, Virginia, pp. 23-44, 1996.
6. J. Thomas, Cross-Cultural Pragmatic Failure, *Applied Linguistics, 3,* pp. 91-112, 1983.
7. R. D. Albert, Polycultural Perspectives on Organizational Communication, *Management Communication Quarterly, 6*:1, pp. 74-84, 1992.

8. T. C. Atkinson, *Merriam-Webster's Guide to International Business Communications,* Merriam-Webster, Inc., Springfield, Massachusetts, 1994.
9. S. Jones et al., *Developing International User Information,* Digital Press, Bedford, Massachusetts, 1992.
10. C. K. Merrill and M. Shanoski, Internationalizing Online Information, in *Proceedings of SIGDOC '92,* Association for Computing Machinery, New York, pp. 19-25, 1992.
11. K. Potosnak, Do Icons Make User Interfaces Easier to Use? *Institute for Electrical and Electronics Engineers Software,* pp. 97-99, May 1988.
12. A. H. Bell, W. T. Dillon, and H. Becker, German Memo and Letter Style, *Journal of Business and Technical Communication, 9*:2, pp. 219-227, 1995.
13. I. Varner, A Comparison of American and French Business Correspondence, *Journal of Business Communication, 25*:4, pp. 55-65, 1988.
14. H. Yli-Jokipii, *Requests in Professional Discourse: A Cross-Cultural Study of British, American, and Finnish Business Writing,* Annales Academiae Scientiarum Fennicae. Dissertationes Humanarum Litterarum 71. Suomalainen Tiedeakatemia, Helsinki, 1994.
15. S. Jenkins and J. Hinds, Business Letter Writing: English, French, and Japanese, *TESOL Quarterly, 21*:2, pp. 327-354, 1987.
16. I. Varner and L. Beamer, *Intercultural Communication in the Global Workplace,* Irwin, Chicago, Illinois, 1995.
17. D. Victor, *International Business Communication,* HarperCollins, New York, 1992.
18. R. Shuter and R. L. Wiseman, Communication in Multinational Organizations: Conceptual, Theoretical, and Practical Issues, in *Communicating in Multinational Organizations,* International and Intercultural Communication Annual 18, R. L. Wiseman and R. Shuter (eds.), Sage, Thousand Oaks, California, pp. 3-11, 1994.
19. N. L. Hoft, *International Technical Communication,* John Wiley & Sons, New York, 1995.
20. W. Horton, The Almost Universal Language: Graphics for International Documents, *Technical Communication, 40,* pp. 682-693, 1993.
21. E. Hall and M. R. Hall, *Understanding Cultural Differences,* Intercultural Press, Yarmouth, Maine, 1990.
22. L. K. Grove, Signs of the Times: Graphics for International Audiences, in *Proceedings of the 1989 International Professional Communication Conference,* Institute for Electrical and Electronics Engineers, New York, pp. 137-141, 1989.
23. R. W. Hartshorn, Writing for International Markets, in *Proceedings of the 37th Technical Writers Institute,* Rensselaer Polytechnic Institute, Troy, New York, pp. 129-135, 1989.
24. J. Rochester, Visual Aids for a Foreign Audience, in *Record of the International Professional Communication Conference,* Institute for Electrical and Electronics Engineers, New York, pp. 533-537, 1992.
25. K. Shriver, Document Design from 1980 to 1989: Challenges That Remain, *Technical Communication, 36,* pp. 316-339, 1989.
26. D. B. Felker, *Document Design: A Review of the Relevant Research,* American Institute of Research, Washington, D.C., 1980.
27. D. B. Felker et al., *Guidelines for Document Designers,* American Institute of Research, Washington, D.C., 1981.

28. L. Flower, J. R. Hayes, and H. Swartz, Revising Functional Documents: The Scenario Principle, in *New Essays in Technical and Scientific Communication: Research, Theory, Practice,* P. Anderson, R. J. Brockmann, and C. Miller (eds.), Baywood, Amityville, New York, pp. 41-58, 1983.
29. L. Faigley and S. Witte, Topical Focus in Technical Writing, in *New Essays in Technical and Scientific Communication: Research, Theory, Practice,* P. Anderson, R. J. Brockmann, and C. Miller (eds.), Baywood, Amityville, New York, pp. 59-70, 1983.
30. J. Selzer, What Constitutes a "Readable" Technical Style? in *New Essays in Technical and Scientific Communication: Research, Theory, Practice,* P. Anderson, R. J. Brockmann, and C. Miller (eds.), Baywood, Amityville, New York, 1983.
31. T. N. Huckin, A Cognitive Approach to Readability, in *New Essays in Technical and Scientific Communication: Research, Theory, Practice,* P. Anderson, R. J. Brockmann, and C. Miller (eds.), Baywood, Amityville, New York, pp. 90-110, 1983.
32. W. H. Motes, C. B. Hilton, and J. S. Fielden, Reactions to Lexical, Syntactical, and Text Layout Variations of a Print Advertisement, *Journal of Business and Technical Communication, 6*:2, pp. 200-223, 1992.
33. P. Benson and R. Burnett, Hand-Made Pages: Information Design with Pen and Mouse, in *Studies in Technical Communication,* B. Sims (ed.), University of North Texas Press, Denton, Texas, pp. 1-19, 1991.
34. P. Benson and R. Burnett, The Shaping of Written Communication, in *Studies in Technical Communication,* B. Sims (ed.), University of North Texas Press, Denton, Texas, 1991.
35. P. J. Benson, Writing Visually: Design Considerations in Technical Publications, *Technical Communication, 32,* pp. 35-39, 1985.
36. A. H. Duin and P. Penn, Identifying the Features That Make Expository Text More Comprehensible, *Contemporary Issues in Reading, 2,* pp. 51-57, 1987.
37. J. Redish, Reading to Learn to Do, *Technical Writing Teacher, 15,* pp. 223-233, 1988.
38. R. C. Parker, *Looking Good in Print* (3rd Edition), also CD-ROM edition, 1996, Ventana Press, Chapel Hill, North Carolina, 1993, 1996.
39. R. C. Parker, *The Makeover Book: 101 Design Solutions* (2nd Edition), with J. Grossmann, Ventana Press, Chapel Hill, North Carolina, 1989, 1995.
40. T. Weiss, 'The Gods Must Be Crazy': The Challenge of the Intercultural, *Journal of Business and Technical Communication, 7,* pp. 196-217, 1993.
41. D. C. Barnlund, *Communicative Strategies of Japanese and Americans: Images and Realities,* Wadsworth, Belmont, California, 1989.
42. G. Hofstede, *Culture's Consequences,* Sage, Thousand Oaks, California, 1984.
43. G. Hofstede, *Cultures and Organizations: Software of the Mind,* McGraw-Hill, London, 1991.
44. G. Hofstede, National Cultures in Four Dimensions, *International Studies of Management and Organizations, 13,* pp. 46-74, 1983.
45. F. Trompenaars, *Riding the Waves of Culture,* The Economist Books, London, 1993.
46. *Abuses within the Mexican Political, Regulatory and Judicial Systems and Implications for the North American Free Trade Agreement: Hearing before the Committee on Banking, Finance and Urban Affairs,* House of Representatives, One Hundred Third

Congress, November 8, 1993, U.S. Government Printing Office, Washington, D.C., 74.B 22/1L103-93, 1993.

47. *US Trade Policy and NAFTA: Hearing before the Commission on Finance,* U.S. Senate, 102nd Congress, March 9, 1993, U.S. Government Printing Office, Washington, D.C., 1993.
48. L. Hoecklin, *Managing Cultural Differences: Strategies for Competitive Advantage,* Addison-Wesley, Wokingham, England, 1995.
49. R. Mead, *Cross-Cultural Management Communication,* Wiley, New York, 1990
50. M. O'Hara-Devereaux and R. Johansen, *Globalwork: Bridging Distance, Culture, and Time,* Jossey-Bass, San Francisco, California, 1994.
51. T. J. Peters and R. H. Waterman, Jr., *In Search of Excellence: Lessons from America's Best-Run Companies,* Warner, New York, 1982.
52. A. Taylor, The Man Who Put Honda Back on Track, *Fortune,* pp. 92-100, September 9, 1996.
53. R. T. Moran and J. Abbott, *NAFTA: Managing the Cultural Differences,* Gulf, Houston, Texas, 1994.
54. P. Drucker, *Management: Tasks, Responsibilities, Practices,* Harper and Row, New York, 1974.

CHAPTER 11

Visual Elements in Cross-Cultural Technical Communication: Recognition and Comprehension as a Function of Cultural Conventions

DEBORAH S. BOSLEY

Imagine this hypothetical situation: ABC Security Systems Inc. spends thousands of dollars, four months of company time, and the expertise of two technical communicators developing documentation for their new security software products. The documents are shipped to Jordan, and within two weeks of delivery, ABC receives a phone call from one of their biggest clients indicating that a translation error caused this client much embarrassment. In a panic, the technical communicators review the documentation and find that the text contains nothing that would be insulting to Jordanian culture. The problem, however, is not located in the text. Because part of the ABC logo is a "sheriff's star," the writers created a cover that contained a circle of six-pointed stars. The Jordanian client interpreted that to mean that ABC was pro-Israel.

Thus, the problem in translation that caused ABC to lose money, time, and respect was not a problem in the text, but rather in the visuals that accompanied that text. The technical communicators were so concerned about making the text translatable that they paid little attention to the cultural impact of the visuals.

Examples of this type abound—international technical communications has typically emphasized the text to the neglect of the visuals. For example, guidelines reported by J. Dana and developed by the Los Alamos National Laboratory (LANL) for preparing documents for export, instruct technical communicators to ". . . avoid idioms and jargon; avoid acronyms, abbreviations, and initialisms; avoid use of one term to name two different concepts; use the active voice; use simple tenses; use clear word order; and avoid long sentences that are compound/complex" [1, p. 148]. The guidelines make no mention of graphic considerations.

Research in anthropology, cognitive psychology, linguistics, and writing theory has identified several factors that affect the way readers read and interpret texts, including cultural differences in processing graphics [2, p. 274]. However, research in technical communication has tended to overlook the importance of visuals in creating international documents, and teachers may be overlooking this important element in educating students. Additionally, technical communication faculty and practitioners have placed less emphasis on the need to create visuals suited for international audiences partially due to the widespread conception that visuals carry fewer cultural connotations than text. Many believe that we can create a neutral visual language that could be used as a replacement for much technical text. As S. Jones et al. suggest in *Developing International User Information,* "[a] well-designed illustration is international; it can provide useful information to customers worldwide without being extensively redesigned, redrawn, or otherwise altered" [3, p. 81].

This chapter addresses the issue of a neutral, visual language appropriate for international documentation by offering a myriad of examples that illustrate attempts to create internationally viable signs and icons and that will sensitize readers to erroneous assumptions about graphics, by discussing the theoretical constructs that inform our thinking about international visuals, and, finally, by suggesting guidelines and assignments that technical communication teachers can use to increase their students' cultural awareness and sensitivity to visual representations of data.

ATTEMPTS TO CREATE INTERNATIONAL SYMBOLS

As technical documentation expands to fill needs throughout the world, the development of a universal, neutral visual language has implications for technical communicators. Books by H. Dreyfuss and E. S. Helfman attest to the attempts to create international symbols as easily recognizable from China to Canada. Such a neutral, visual language would save companies thousands of dollars and hundreds of hours of translation time. Indeed, the international changes in the workplace create a great need for translation [4, 5]. However, despite claims about the universal appeal of graphic symbols, many attempts to create international symbols and icons have failed [6]. For instance, in a study of six international, visual referents developed to aid international travelers, Zwaga and Easterby located the following number of international symbol variants for each reference being investigated [7, p. 281]:

- Drinking water 15
- Information 29
- Stairs 16
- Taxi 23
- Toilets 35
- Waiting room 8

In 1981, Zwaga and Easterby conducted another study of a set of public information symbols for railway stations. Their study included 400 respondents who were to determine the meaning of twenty-nine public information symbols. The results indicate that only the signs for "telephone" and "toilet" are "almost always correctly chosen" by the respondents [6, p. 290]. Figure 1 illustrates those symbols used by Zwaga and Easterby.

In 1973, the International Organisation for Standardisation (ISO) "initiated . . . the development of an international standard for public information symbols" [7, p. 278]. ISO intended to standardize graphic symbols to account for differences among nations. Currently ISO is working both on developing standards that attempt to increase the possibility of internationalizing symbols in technical fields, and a sub-group of ISO is "considering standards for signs of the user interface" [8, p. 258].

Despite such well-intentioned attempts to create or to determine culturally neutral symbols, it appears that only in specialized fields such as medicine or engineering can such symbols become more "neutral" or "universal." Technical conventions can sometimes create specialized discourse communities or industry standards that transcend cultural and national boundaries. Indeed, according to C. Kostelnick, "[m]any visual discourse communities owe their existence to conventional codes . . . that members of these communities acquire through extensive training, experience, and interaction . . ." [9, p. 185]. Symbols used in circuit diagrams to show logic gates, electrical switches, and other electrical parts, like their counterparts in other fields in engineering and science, may be capable of making Transatlantic leaps across cultural communities: most icons or symbols, however, are not.

Many attempts fail because of the lack of knowledge about how other cultures respond to symbols. K. Maitra and D. Goswami in "Responses of American readers to visual aspects of a mid-sized Japanese company's annual report: A case study" illustrate differences between how North American, English-speaking cultures and the Japanese interpret and use graphics [10, p. 198]. They indicate that because the Japanese are extremely visually oriented, they access information more easily through the use of and preference for visual information. "The

Referent	Symbol	% correct	Referent	Symbol	% correct
Entrance	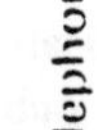	56	Telephone		94
Exit	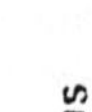	66	Bus		62
Information office		84	Taxi		87
Buffet	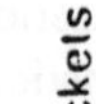	33	Tickets		55
Restaurant	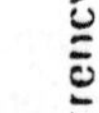	55	Currency exchange		81
Food vending machines	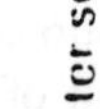	64	Porter service		70
Toilet		99	Station parking lot		75
Left luggage	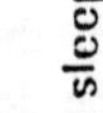	34	Car sleeper		74
Luggage dispatch		19	Car rental		75

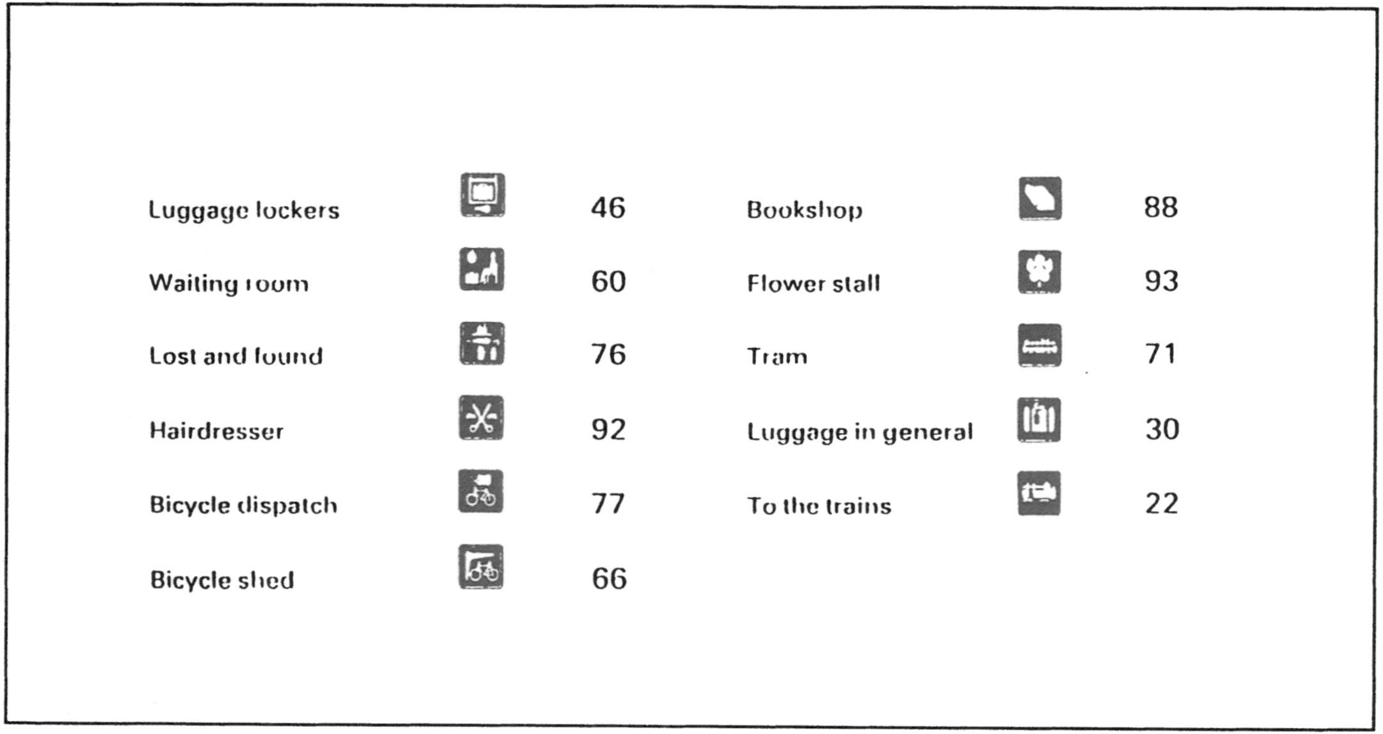

Figure 1. H. Zwaga and R. Easterby, Developing Effective Symbols for Public Information, in *Information Design: The Design and Evaluation of Signs and Printed Material,* R. Easterby and H. Zwaga (eds.), John Wiley & Sons Ltd., Chichester, England, p. 290, 1984.

Japanese can easily process more complicated information through a visual than can people from other countries. Indeed, they prefer visual representation of any information over verbal representation." Such distinctions are apparent in that "[c]ontrary to the Japanese idea of maintaining aesthetics and ambiguity in the use of visuals in cross-cultural documents, the [North] American concept of internationalizing graphics aims at clarity and culture neutrality" [10, p. 198].

CONTRASTING VIEWS ON DESIGNING INTERNATIONAL VISUALS

These attempts to create international icons represent one of the two divergent viewpoints presented by C. Kostelnick in his article "Cultural adaptation and information design: Two contrasting views": a global (universal) approach and a culture-focused approach [9, p. 183]. Table 1 illustrates the differences between the global (universal) and the culture-focused approach to visual design.

The global approach posits that the cognitive functions of perception, because they do not vary from culture to culture, allow technical communicators to create universal recognition of icons and symbols. A global focus assumes that visual language is, like verbal language, a transparent form that allows meaning to reside within the visual representation itself. On the other hand, the culture-focused approach posits that icons and symbols can be understood only in the social, historical, and psychological context of the user's culture.

Table 1. Global versus Culture-Focused Approaches to Visual Design

Global	Culture-Focused
designers aim to create an objective, universal, visual language that requires little attention to cultural context	designers understand that visual language is largely a social construct that proceeds from experience, varies across cultural groups, and requires sensitivity to context
functions across international boundaries	functions only within a given cultural context
can be designed for international audiences by understanding the cognitive processes of vision which, by activating conceptual capabilities of the eye and brain, remain the same from culture to culture	can be designed only by understanding the historical, social, and experiential elements in each culture: emphasis on translation with assistance from user's culture
anchored firmly in perception theory and research and in twentieth-century modernism	anchored in contemporary theories ranging from social constructionism to postmodernism

This culture-focused view provides us with a better likelihood of understanding since "the communicability of any representation depends on a shared context between sender and receiver" [11, p. 188]. That is, just as technical communicators and educators recognize the influence that cultural context has on language, document designers should shift the emphasis away from the logical positivists' perspective of creating a neutral, visual language to a social constructionist's perspective. However, doing so demands that we understand the cultural elements that influence how users read, comprehend, and respond to texts and visuals from country to country or region to region.

SIGNS, SYMBOLS, AND ICONS AS CULTURAL ARTIFACTS

Before proceeding further, a definition of signs and icons may help reinforce the difficulties technical communicators encounter in attempting to create culturally neutral visuals whether such visuals are displayed in hardcopy, in online documentation, or through graphical-user-interfaces (GUI).

A sign or an icon "denotes its object by virtues of its own likeness to or resemblance of that object, on the basis of some quality of characteristic inherent in the icon itself" [11, pp. 172-173]. For example, a drawing of a flame—to denote the concept of "fire"—may be used on a manufacturing product label to warn the user that the product is flammable.

However, signs and icons only have meanings if the observer understands the association. That is, the user or reader must be able to interpret "the meaning of a sign by identifying the sign object and grasping the significance of the object" [11, p. 172]. The metaphor of a "desktop," which is found in many GUI environments, uses an iconic form of representation through such visuals as the folders, trash cans, mailboxes, in or out baskets, etc. The assumption is that when users see these icons, they will know immediately what function that icon represents.

Perhaps one of Macintosh's greatest accomplishments was to bring the word "icon" into the general vernacular. Their graphic-user-interface (GUI) approach produced a series of icons that seemed at once both simple and universal. As a visual shorthand for conveying information, icons seemed to offer the following advantages for communicating with international users. Signs and icons can

- communicate some messages more effectively than text, notably safety warnings or cautions
- create the perception of an international product
- replace technical terms that are difficult to translate
- fit into spaces too small for text [3, p. 93].

However, as K. Mullet and D. Sano point out in *Designing Visual Interfaces,* "effective visual representations for international audiences should be based on aspects of the sign object that are truly universal within the target population" [11, p. 191].

Despite such admonitions, many major companies fail to consider the influence of culture of users on their ability to recognize the significance of icons. For example, Macintosh created the "file" icon under the assumption that it would be easily and universally recognizable. It isn't. A. Marcus points out that in North American offices documents are stored in stiff folders and placed horizontally in file cabinets or desk drawers. On the other hand, in Japan and many European countries, documents "are stored in cardboard box-like containers on shelves, with the vertical paper sheets punched and held in place by rings . . . [and] are pulled off the shelves using a small finger hole located in the vertical face of the box" [8, p. 259]. Figure 2 shows a more appropriate icon for a Japanese audience.

Even the most familiar objects are not necessarily understood by international audiences. The mailbox, for instance, is commonly used to represent electronic mail. The common "country" version of a mailbox, while recognizable to North American users, may be better represented internationally through the icons in Figure 3. In their attempt to create a set of useful, recognizable icons, Macintosh

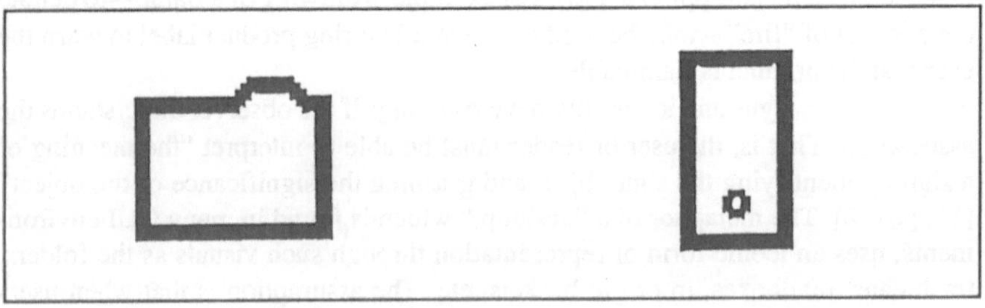

Figure 2. A. Marcus, Icon and Symbol Design Issues for Graphical User Interfaces, in *International User Interfaces,* E. M. del Galdo and J. Nielsen (eds.), John Wiley & Sons Ltd., New York, p. 259, 1996.

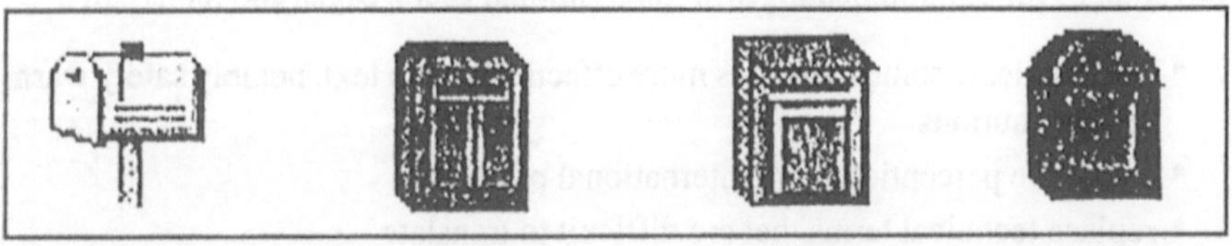

Figure 3. A. Marcus, Icon and Symbol Design Issues for Graphical User Interfaces, in *International User Interfaces,* E. M. del Galdo and J. Nielsen (eds.), John Wiley & Sons Ltd., New York, p. 260, 1996.

neglected to consider that icons generally are dominated by cultural, historical, professional, and social conventions.

Symbols, on the other hand, are not necessarily "simplified graphic representations of a particular object or situation [because] understanding a symbol often calls for learning: the meaning of a symbol cannot simply be inferred" [3, p. 93]. Symbols may be abstractions relevant to particular industries or professions. Figure 4 shows electronic symbols used internationally [12].

Symbols also may resonate on a sociological and psychological level because they carry with them deep levels of cultural understanding. For the Japanese, for example, the symbol of the "cherry tree" carries with it a haiku tradition, resonances of religion and spirituality, and sense of the country's relationship with nature. In the United States, the "cherry tree" evokes images of George Washington and symbolizes the wisdom of telling the truth. Neither culture would interpret the same image of the "cherry tree" in the same way. In this sense, symbols do not readily convey the intended meaning until either a specific message has been assigned to it or people have learned the association.

UNCOVERING CULTURAL ASSUMPTIONS

Technical communicators who create visual elements within documents prepared for translation or export or technical communication professors who are preparing their students for international audiences should understand many of the common visual elements about which we hold graphic assumptions. The following story exemplifies how a common graphic sign was used (with disastrous results) assuming that all readers would interpret it the same way.

A common sign in North American postal and shipping industry system is the use of a broken glass to symbolize "fragile." E. S. Helfman, in *Signs and Symbols Around the World,* tells the story of a large wooden crate being shipped to India [5, pp. 165-166]. The crate contained glassware and a large symbol of a broken glass. When the crate arrived, the dockworker looked at the label and assumed that the broken glass represented the contents. Assuming that it must be full of broken glass (never mind the question of why someone would ship broken glass), he picked up the crate and threw it onto the back of a truck. The packers had assumed that the symbol of a broken glass was a universal symbol.

The following elements of designing visuals, punctuation marks, religious symbols, colors, people, and gestures reveal our ethnocentric assumptions about what is universal.

Punctuation Marks

Punctuation marks, just like words, are language specific. For example, in North America we might use a ? to represent a need for "information." This mark, however, has meaning only in American or Western European languages. The

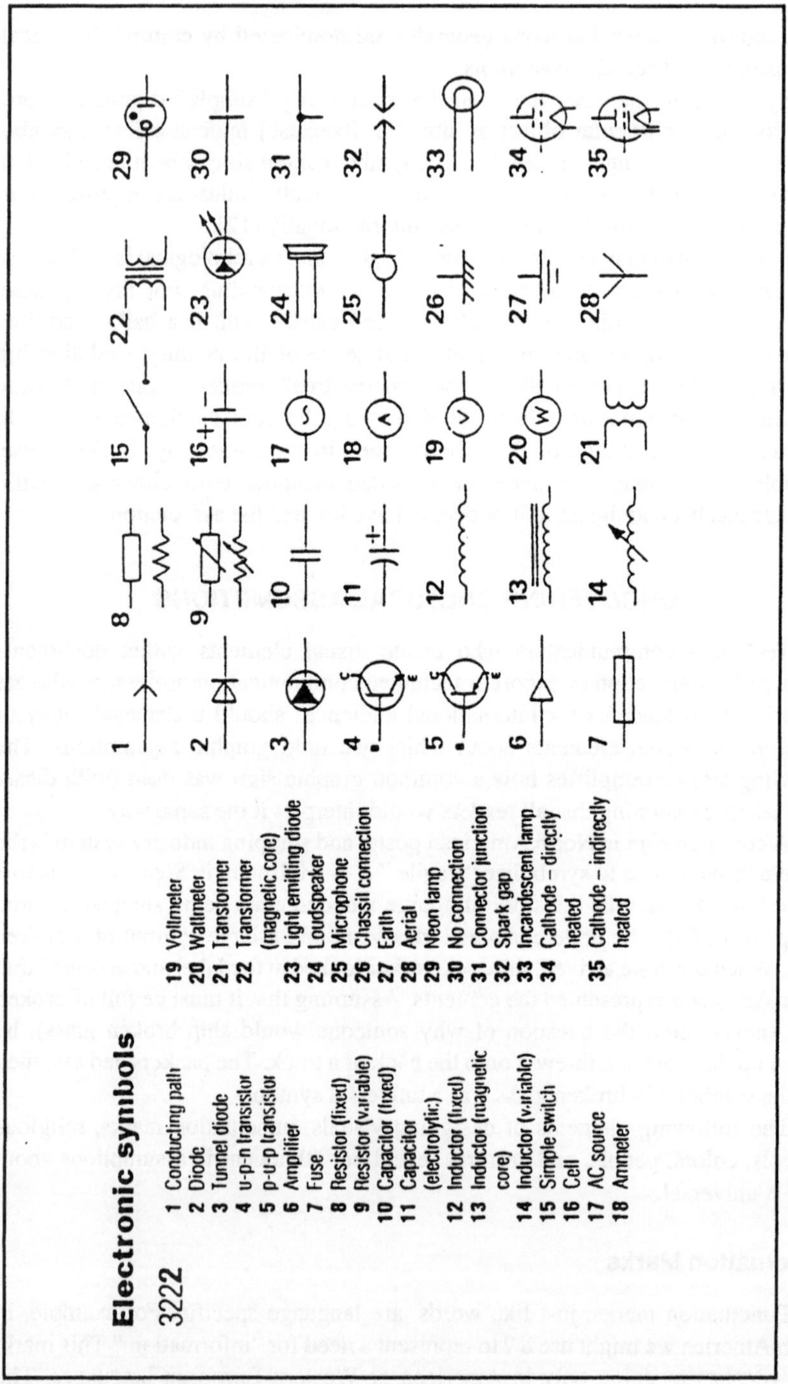

Figure 4. *The Macmillan Visual Desk Reference*, Macmillan Publishing Co., New York, #3222, 1993.

ISO 1973 test series conducted by Easterby and Zwaga concluded that the use of a question mark to represent "information" was only understood by 48 percent of the population tested [7, p. 284]. Our question mark is not a universal sign for "questions" or for "information."

Religious Symbols

Religious symbols that carry non-religious meaning have long been used in North America where it is common to use a + as a symbol for "First Aid" or for "Hospital." But to Muslims, a cross of any kind represents Christianity. Thus, in Muslim countries a First Aid symbol is a crescent. In this case, our predominant religious symbol is assumed to have non-religious connotations.

Colors

The use of color in technical documents can telegraph a variety of meanings. In North America, Europe, and Japan it is common for the color red to be used as "Warning" or "Danger." In China, however, red symbolizes "joy." In Europe and North America, blue generally has a positive connotation; in Japan, the color represents villainy. In Europe and North America, yellow represents caution or cowardice; in Arabic countries, yellow generally means fertility or strength. Suggestions for designing promotional materials for English as a Second Language program include not using black, gray, or yellow as a primary color [13, p. 177]. A blue ribbon means first prize in the United States, but second prize in the United Kingdom where first prize is indicated by a red ribbon. Thus, assumptions about the meaning and uses of color are not universal.

People

The use of people and body parts can be problematic. For example, nudity in advertising generally is acceptable in Europe, but not in North America. One technical manual, in attempting to show the ease of running a particular software program, showed the user with his feet up on a desk. Problems arose when this manual was shipped to the Middle East because in those cultures, showing the soles of one's feet or shoes is an insult. Even a smile (the ubiquitous smiley faces) may have different connotations: smiling can be a sign of joy in Japan or can be used to hide displeasure; in some Asian cultures smiling may be a sign of weakness.

In general, representations of people are unlikely to become international or universal because how men and women dress, what roles they play, and what revealed body parts are insulting change from culture to culture.

Gestures

Technical communicators often use gestures as a means of indicating direction (pointed finger), danger (extended palm as "stop"), and success (the "ok" sign). It is not uncommon to see many of these gestures imbedded within a set of instructions or a user manual as a means of indicating caution or as positive motivators (see Figure 5).

However, such gestures carry with them cultural connotations. In 1992, when passing a group of protestors in Canberra, Australia, President George Bush held up his fingers in a "victory" sign. He failed to understand that this same gesture in Australia is the equivalent of holding up the middle finger. The North American

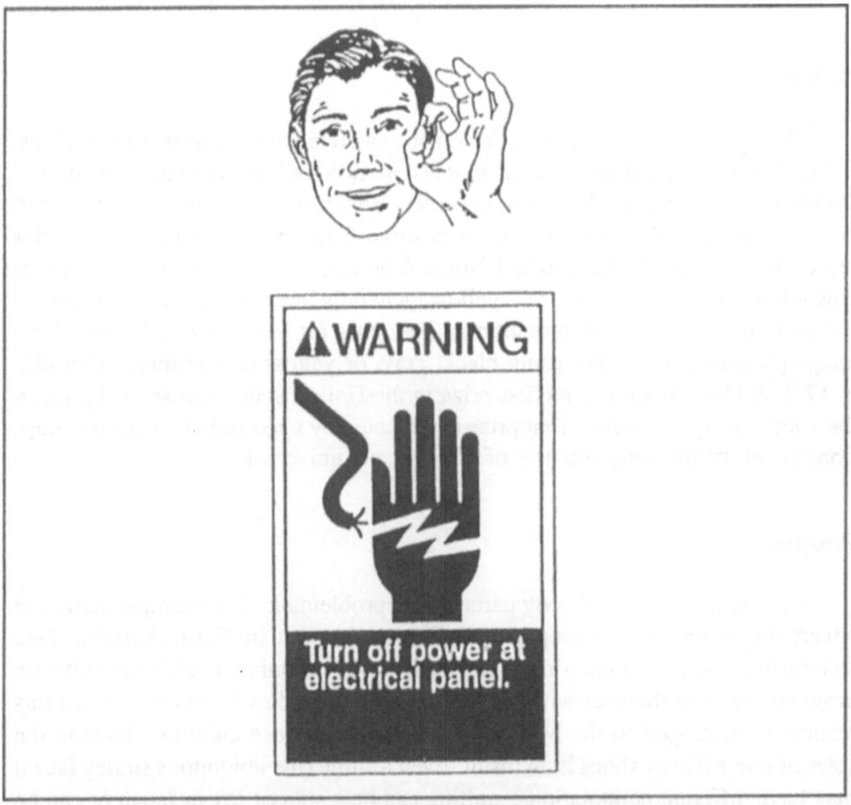

Figure 5. F. Robinett and A. Hughes, Visual Alerts to Machinery Hazards: A Design Case Study, in *Information Design: The Design and Evaluation of Signs and Printed Material*, R. Easterby and H. Zwaga (eds.), John Wiley & Sons Ltd., Chichester, England, p. 409, 1984.

gesture that means "OK," which consists of forming a circle with the thumb and index finger, can mean "worthless" to the French, "money" to the Japanese, "male homosexual" in Malta, and is a general sexual insult in many other parts of the world [14, pp. 118-119]. Table 2 indicates additional international implications of gestures around the world.

Thus, a manual that contained a pointed finger to indicate "turn the page," might offend someone in Venezuela. Or, for instance, if a writer were preparing a manual for export to Honduras and wanted to warn the reader about possible computer error, or export a set of instructions for machinery and wanted to warn the reader about possible dangers, imbedding a picture of a person touching a finger below the eye to indicate caution would be more appropriate.

Even within the United States, with our increasing international population, we need to recognize the affect that hand gestures can have. The National Motorists Association, for example, attempted to create seven hand signals that all drivers could use to indicate the need for roadside assistance. Figure 6 illustrates those signals that could cause, at the least, considerable laughter among our international residents.

Reading Habits

Whether a culture reads right to left or left to right or vertically influences how graphics are sequenced. For example, in Israel *The New York Times* is read from the back to the front mirroring the right to left movement of the Hebrew language. People whose language is read from right to left are more likely to view visuals counterclockwise rather than the North American tendency to view visuals clockwise. Japanese, Chinese, or Korean readers, whose ideographic language runs vertically, prefer technical manuals that are designed to be read both vertically and right to left. Thus, such knowledge of reading directionality would help determine design and sequencing. Figure 7 illustrates a safety warning for those

Table 2. International Implications of Gestures

U.S. Gestures		Other Countries' Interpretation	
Head nodding	Up and down	Bulgaria	No
Left hand	Showing palm, shaking hands	Muslim countries	Dirty, unclean
Index finger	Pointing to others	Venezuela, Sri Lanka	Rude
Index finger	Pointing to self	Germany, Netherlands, Switzerland	Rude
Index/thumb circle	Circular "ok"	Brazil, Paraguay, Singapore, Russia	Sexual connotations
Ankle and leg	Crossing over knee	Indonesia and Syria	Rude

PULL OVER FOR PROBLEM

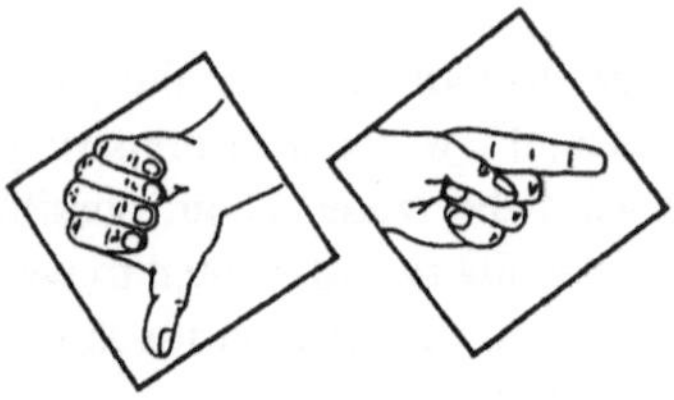

You come across a vehicle about to have a flat, or lose luggage from an outside rack or litter the highway with skis, bicycles, or furniture. The "Pull Over For Problem" signal helps you alert the other driver.

To alert another motorist of a problem with his or her vehicle, first point in the direction of the problem (up for lose roof rack, back for trailer problem, etc.), then signal "thumbs down."

If you receive this signal from another motorist, pull over and check your vehicle.

I UNDERSTAND
(Thank You, I Understand)

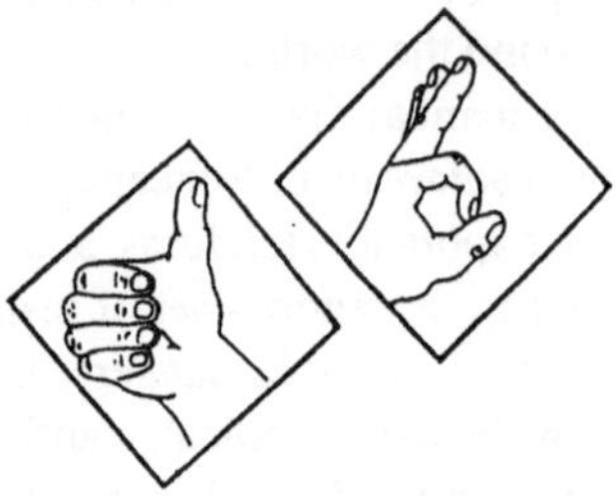

To acknowledge another motorist's signal . . . an apology, light problem, or problem pull-over signal, for example, or to thank another driver for a courtesy, use the well-understood "thumbs up" or "OK."

Figure 6. Seven Sensible Signals, National Motorists Association, Wisconsin.

operating machinery [15, p. 265]. Note the confusion that might occur in this safety cartoon if this visual were used with a reader who accessed information from right to left.

The previous examples illustrate merely a few of literally thousands of examples of poorly conceived signs and icons intended for international usage [e.g., 11, 16-18].

If we recognize that creating neutral visuals (except in specific technical fields) may be impossible, what concepts can we use to teach students how to create visuals for international audiences? One of the most viable methods for studying and understanding cultural differences that affect document design is E. T. Hall's discussion of high-context and low-context cultures [19]. Although commonly used to explain issues of text development, an understanding of

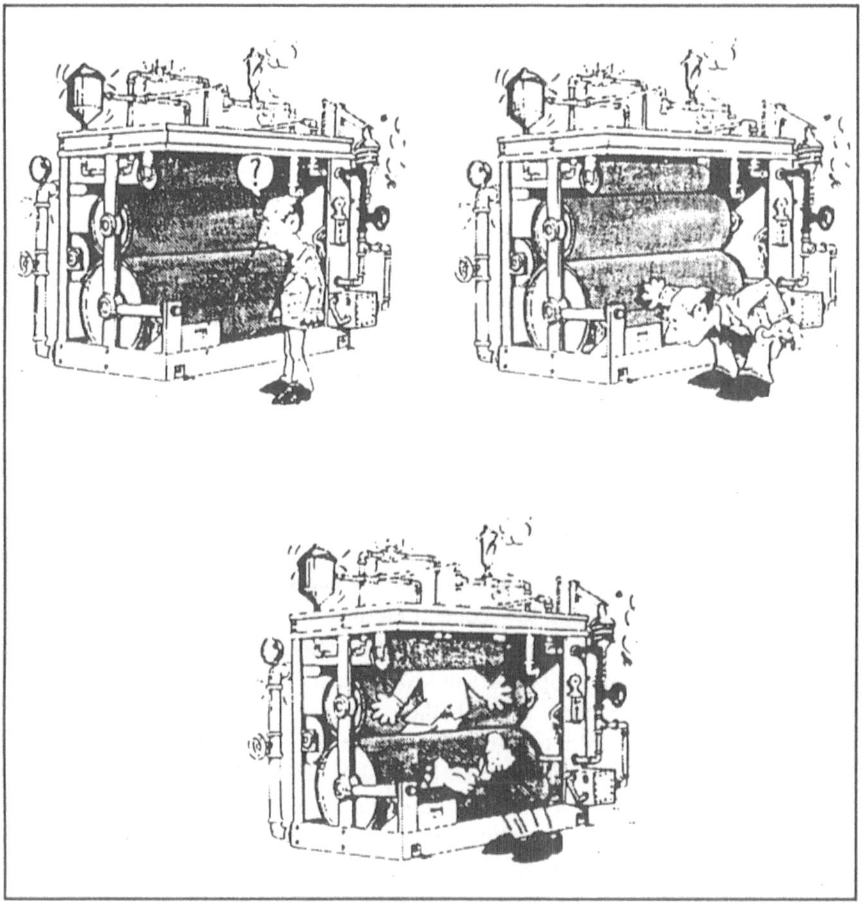

Figure 7. F. Robinett and A. Hughes, Visual Alerts to Machinery Hazards: A Design Case Study, in *Information Design: The Design and Evaluation of Signs and Printed Material*, R. Easterby and H. Zwaga (eds.), John Wiley & Sons Ltd., Chichester, England, p. 406, 1984.

high- and low-context cultures also contributes to our design and adoption of visual elements.

HIGH-CONTEXT AND LOW-CONTEXT INFLUENCES ON VISUALS

Hall, in *The Dance of Life,* delineates cultures by whether they are high context or low context. High context or low context refers to the amount of information that is in a given communication as a function of the context in which it occurs. A high-contexted communication is one in which most of the meaning is in the context while very little is in the transmitted message. A low-context communication is similar to interacting with a computer—if the information is not explicitly stated . . . the meaning is distorted [20, p. 229].

In high-context cultures, people share similar ethnic, religious, and educational backgrounds. They tend not only to have the same knowledge of the world, but also to share attitudes, feelings, and values. Technical writers in high-context cultures do not provide a great deal of background information or burden the reader with too much detail precisely spelled out. "In fact, it is considered insulting to the [Japanese] reader to include too many details as if the reader were incapable of filling in the gaps" [2, p. 275]. In the United States, which is a "low-context" culture, regional, ethnic, educational, and religious differences have created a country in which its inhabitants may have few shared knowledge, values, or attitudes. Table 3 illustrates this continuum as readers move from low- to high-context cultures.

In high-context cultures, a speaker relies on the listener to fill in the context for much of the information. More of the message is left unspoken and is accessed through nonverbal cues and interpretations of what is meant rather than of what is said. In the United States, a low-context culture, speakers are more specific and direct. The speaker does not rely on context to convey meaning; and the listener does not need to interpret. Speakers in low-context cultures also tend to talk more frequently, provide context through their spoken words, and be more direct. Workers and students from high-context cultures tend to speak far less frequently and rely on context for meaning. D. S. Bosley [21] and E. Thrush [2], among others, have given advice for using high- and low-context distinctions as means for writing technical documentation. For example, they suggest that when writing for audiences in low-context cultures, writers must create and explain the context

Table 3. Meaning as a Function of High- and Low-Context Cultures

Meaning Relies on Implicitly Stated Information in High-Context Cultures
Japanese, Chinese, Korean
Arabic
Latin American
Italian
British
French
North American
Scandinavian
German
Swiss-German
Meaning Relies on Explicitly Stated Information in Low-Context Cultures

of the document (usually called the background), give a tremendous amount of detail (analysis), and be direct in their conclusions and recommendations. When writing for readers in high-context cultures, writers may omit a great deal of background information and may be more vague and ambiguous. Readers in high-context cultures rely on their own internal context to "fill in the gaps."

Although much discussion of high- and low-context cultures focuses on speaking or writing, elements of how readers come to make meaning from visuals for international audiences are shown in Table 4.

As technical communicators come to recognize the influence that high- and low-context cultures have on designing both international texts and visuals, the issues of localization and internationalization can provide us with some solutions.

THE LOCALIZATION VERSUS THE INTERNATIONALIZATION APPROACH TO VISUAL DESIGN

N. L. Hoft defines localization as "the process of creating or adapting an information product for use in a specific target country or specific target market" [16, p. 11]. Two forms of localization are relevant for this discussion: general localization and radical localization. General localization means translating differences inherent in "language, currency formats, date, and time formats" [16, p. 12]. For example, a North American company that produces a financial software product could prepare the documentation for export to Great Britain by changing dollar signs to pound signs. Radical localization involves creating icons or symbols that incorporate cultural differences that affect the way users think, feel, act, and learn. In considering radical localization, technical communicators have to consider and use the information they have about the deeper psychological influences that affect how users create meaning.

Hoft further illustrates her discussion of localization with the analogy of an iceberg. The upper one-quarter identifies typical general localization influences such as linguistics, technology, environments, and economies. The lower

Table 4. The Influence of High- and Low-Context Cultures on the Use of Visuals in Document Design

In High-Context Culture	In Low-Context Culture
More white space	More information on visual
Less text within visual	Detailed explanations within visual
Let the image "speak for itself"	In the document, "surround" the visual with detailed explanations and analyses
More abstract visuals	More concrete visuals

three-quarters, which identifies radical localization elements, include the more "unconscious" influences on our lives: e.g., religion, education, social organizations, time, space, collective vs. individual emphases, masculine vs. feminine behaviors, and conceptions of authority [16, p. 52].

Radical localization implies that technical communicators must understand the deep structures of a culture in order to evoke context-situated meanings and responses among users. As E. T. Hall and M. R. Hall wrote in *Understanding Cultural Differences,* "The essence of effective cross-cultural communication has more to do with releasing the right responses than with sending the 'right' messages" [22, p. xiv]. These "right responses" depend on the technical writers' abilities to comprehend many of the deep structures that influence how users will interpret visuals accompanying technical documentation. Given the number of international exports to hundreds of cultures, is it realistic to expect a technical communicator in Charlotte, North Carolina, to understand the deep meanings appropriate to users in Canton? Probably not.

However, technical communicators can look to the process of "internationalization" for a more appropriate and possible resolution to this dilemma. Internationalization is the process of redesigning documentation so that it can be easily localized for exportation. An internationalized information product consists of two components: core information and international variables [16, p. 19]. Core information (or core text) is "information that remains the same, except for minor variations, across an entire family of products, or for the same product used in different operating environments" [3, pp. 18-19]. For example, when preparing instructions on editing functions within a particular word processing system, technical communicators can use the same instructions for all systems in which that word processing system is used. According to Digital, core text "was used in about 30% of the hardware user information sets translated at Digital in 1990. On average, the user information was translated into three languages" [3, p. 30]. By replicating core information, Digital was able to save both time and money and leave much of the localization to on-site, country-specific translators who were more familiar with radical, deep structures of meaning. These international variables that must be localized both for "superficial and for deeper cultural differences . . . including the page design, the writing style, the cultural context of graphics and of examples, . . . units of measurement, and the time, date, and currency formats" [16, p. 19]. Table 5 suggests guidelines for preparing both texts and visuals for an international audience.

GUIDELINES FOR TEACHING INTERNATIONAL GRAPHICS

In order to help students become aware of the need to design or use icons, signs, or symbols that are sensitive to the cultural expectations of a target audience, general guidelines for teaching international graphic awareness include:

Table 5. Guidelines for Producing Cross-Cultural Text and Visuals

Texts	Visuals
• Avoid idioms and jargon	• Avoid English-only symbols such as punctuation marks, Uncle Sam, etc.
• Avoid acronyms, abbreviations	• Avoid religious symbols
• Use terms and words consistently	• Be consistent in labeling elements
• Avoid referring to readers by nationality	• Avoid using people/body parts; use outlines or neutral abstractions
• Use more detail and background information for low-context readers; less verbiage for high-context readers	• Use more white space for low-context readers; more text for high-context readers
• Include a glossary of terms	• Explain meaning of icons/symbols
• Avoid long, complex sentences	• Avoid using complex visuals
• Run usability testing with members of target, international audience	• Test icons and symbols in context with members of target audience
• Use active voice	• Use simple terms to describe elements
• Organize information appropriately for intended international audience. North American readers want brief introductions, overviews of product and its functions, and a step-by-step tutorial, in that order. Japanese prefer to read discrete parts, not the big picture	• Organize visual information for intended, international audience. North American readers tend to read visuals from left-to-right in clockwise rotation; Mid-Eastern cultures read visuals from right-to-left in counterclockwise fashion.

- acknowledging diversity within the classroom: discuss openly the differing cultures within your classroom or within your region by helping students understand different cultural conventions that affect visual design.
- developing a set of assignments to uncover cultural assumptions among students: such assignments can be drawn from real workplace situations, from textbooks in anthropology, international business, and cross-cultural communication.
- discussing common gestures from other cultures: gestures are a good place to begin a discussion of international graphics since differences are so readily available (and often humorous).
- gathering information from other disciplines: anthropology, art history, architecture, graphic design, information design, and business communication are particularly rich areas for examples and theory.
- using international students as editors and advisors: international students can be quite helpful in assessing intended graphic elements as well as providing information about cultural differences between their country and ours. International students tend to be much better versed in cultural differences, while North American students tend to be more ethnocentric

and ascribe behavior differences to personality rather than to cultural influences [23].

- inviting international, corporate employees to classes: companies in your locale are likely to have several international employees who could be resources for cultural distinctions. Because many have gone through corporate training on diversity, they are likely to be well-versed in cultural conventions.
- gathering examples from the workplace: include visuals that would not translate easily to another culture or that could illustrate cultural sensitivity or insensitivity.

ASSIGNMENTS FOR TEACHING INTERNATIONAL GRAPHICS

These guidelines suggest a number of possible assignments that would enable technical communication teachers to uncover graphic assumptions, explain underlying theories, and sensitize students to the issue of international visuals. Assuming you are teaching an introductory course in technical communication, you might consider the following:

- Ask each student to draw a picture of a house. Put students in teams to discuss their drawings. The differences among the houses students are likely to draw can illustrate some cultural conventions.
- Assign teams of three to research the visual, cultural conventions in a particular country. Require them to present their findings using a series of visuals that illustrate the results of their research.
- Search the Internet 1) for examples of graphics created in other countries, 2) to develop e-mail relationships with students in other countries, and 3) to analyze Websites for their assumptions about how users understand visual representations.
- Brainstorm with the class a series of questions they have about cultural differences between native and international workers. Have students identify and interview international employees in local industries and businesses. Such interviews can be written up as memos, reports, or verbal presentations to the entire class.
- Ask international students to discuss their reaction to the visual examples given earlier in this chapter.
- Give teams an assignment to create a set of instructions that include visuals. Ask either international students or international workers to serve as editors pointing out both the textual and the visual elements that might be problematic in their native language.

- Help students find internships in international companies or where international workers are employed.
- Use the following figures to show students the difficulty of designing icons that are recognized by members of different societies. Remove the explanation of the icons until students have a chance to guess their meanings (see Figure 8).

Marcus explains that icons one and two represent Telephone and Post Office. Apparently icon one suggests the mouthpiece and earphone of a telephone receiver while icon two represents a "horn or bugle, which is sometimes used for national post office and telecommunication centers" in Europe [8, p. 269]. Icons three and four represent Plant and Meat Quarantine with the heart suggesting "caring" and the sausage representing meat. Finally, icons five (Archiver) and six (Back to Initial Screen) represent functions of the Ricoh GW2000 workstation [8, p. 270].

THE FUTURE OF VISUAL ELEMENTS IN A GLOBAL ECONOMY

Clearly the impetus for technical communicators to create more international visuals will increase because of their economic impact and our shift to a more visual culture. As companies look for more and more ways to save money on documentation, technical communicators realize that time and money are savings directly related to the use of visuals. Icons take fewer pixels and use less space

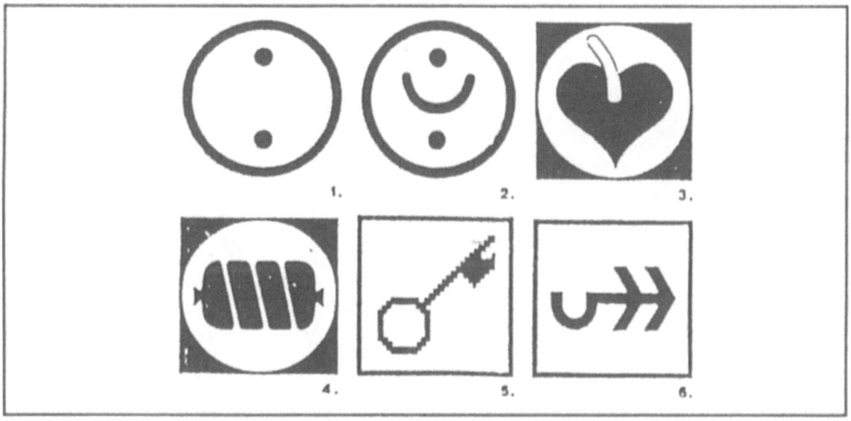

Figure 8. A. Marcus, Icon and Symbol Design Issues for Graphical User Interfaces, in *International User Interfaces,* E. M. del Galdo and J. Nielsen (eds.), John Wiley & Sons Ltd., New York, p. 269, 1996.

than words, thereby packing more information onto pages or screens "saving space in menus, maps, and diagrams" [8, p. 263]. Visuals can be used to provide information quickly in international training situations where time is of the essence. And visuals can symbolize an entire culture. At the beginning of this article we saw how an uninformed technical communicator at ABC Security Systems Inc. failed to recognize the cultural implications of a six-pointed star. R. Wurman shows how to represent cultural implications through his diagram of the route of the Yamnanote Railway in Tokyo. Using the symbol of the yin-yang evokes thousands of years of religion and philosophy and makes the diagram immediately accessible (see Figure 9) [24, p. 268].

Because of the influence of television and computers (among others), users (particularly younger readers) access information more visually than in the past.

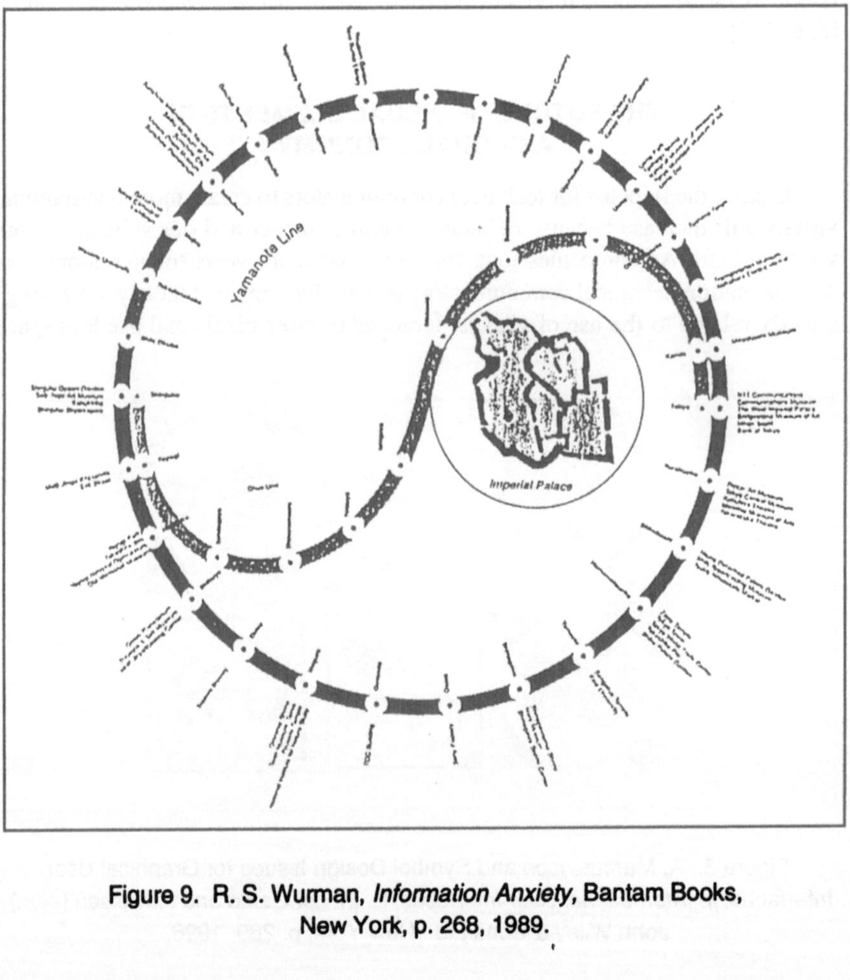

Figure 9. R. S. Wurman, *Information Anxiety*, Bantam Books, New York, p. 268, 1989.

Such a shift from a print-based to a visual-based society demands that technical communicators provide users with more visuals. As Jones et al. indicate, "An entire repertoire of new icons and symbols must be researched, designed, tested, and introduced into the marketplace. There is no system already in existence that may be readily incorporated, unlike letterforms and typefonts" [3, p. 263].

As we respond to the increased needs of international users, the following questions may help guide our research:

- Are the processes for localizing visuals influenced by the gender of the user or the document designer?
- What evidence do we have that signs designated as international symbols are understood by the majority of users?
- What aesthetic considerations should affect the design of international graphics?
- Which icons or symbols are easily recognized internationally? What inherent elements make them universal?

The answers to these (and other) questions will enable us to educate our technical communication students for the world of work in the twenty-first century. And as we continue our increased emphasis on preparing students for an international marketplace and as they learn to prepare technical documents for translation, we are changing the elements of effective technical communication to include an international perspective on both the verbal and the visual.

REFERENCES

1. J. Dana, Addressing a Worldwide Readership Through the Filter of Translation, in *IEEE Transactions on Professional Communication, 3,* pp. 147-150, 1991.
2. E. Thrust, Bridging the Gaps: Technical Communication in an International and Multicultural Society, *Technical Communication Quarterly, 2,* pp. 271-283, 1993.
3. S. Jones et al., *Developing International User Information,* Digital Press, Bedford, Massachusetts, 1993.
4. H. Dreyfuss, *Symbol Sourcebook: An Authoritative Guide to International Graphic Symbols,* McGraw-Hill, New York, 1972.
5. E. S. Helfman, *Signs and Symbols Around the World,* Lothrop, Lee & Shepard Co., Inc., New York, 1967.
6. H. Zwaga and R. Easterby, Developing Effective Symbols for Public Information, in *Information Design: The Design and Evaluation of Signs and Printed Material,* R. Easterby and H. Zwaga (eds.), John Wiley & Sons Ltd., Chichester, England, pp. 277-297, 1984.
7. R. Easterby and H. Zwaga (eds.), *Information Design: The Design and Evaluation of Signs and Printed Materials,* John Wiley & Sons Ltd., Chichester, England, 1984.

8. A. Marcus, Icon and Symbol Design Issues for Graphical User Interfaces, in *International User Interfaces,* E. M. del Galdo and J. Nielsen (eds.), John Wiley & Sons Ltd., New York, pp. 257-270, 1996.
9. C. Kostelnick, Cultural Adaptation and Information Design: Two Contrasting Views, *IEEE Transactions on Professional Communication, 4,* pp. 182-196, 1995.
10. K. Maitra and D. Goswami, Responses of American Readers to Visual Aspects of a Mid-Sized Japanese Company's Annual Report: A Case Study, *IEEE Transactions on Professional Communication, 4,* pp. 197-203, 1995.
11. K. Mullett and D. Sano, *Designing Visual Interfaces,* SunSoft Press, New York, 1995.
12. *The Macmillan Visual Desk Reference,* Macmillan, New York, p. 1712, 1993.
13. F. L. Jenks, Designing and Assessing the Efficacy of ESL Promotional Materials, in *Building Better English Language Programs: Perspectives on Evaluation in ESL,* M. C. Pennington (ed.), NAFSA/Association of International Educators, Washington, D.C., pp. 239-252, 1991.
14. D. Morris, *Bodytalk: The Meaning of Human Gestures,* Crown Trade Paperbacks, New York, 1994.
15. F. Robinett and A. Hughes, Visual Alerts to Machinery Hazards: A Design Case Study, in *Information Design: The Design and Evaluation of Signs and Printed Material,* R. Easterby and H. Zwaga (eds.), John Wiley & Sons Ltd., Chichester, England, p. 265, 1984.
16. N. L. Hoft, *International Technical Communication: How to Export Information About High Technology,* John Wiley & Sons Ltd., New York, 1995.
17. D. A. Ricks, *Big Business Blunders: Mistakes in Multinational Marketing,* Dow Jones-Irwin, Homewood, Illinois, pp. 76-77, 1983.
18. E. M. del Galdo and J. Nielsen, *International User Interfaces,* John Wiley & Sons Ltd., New York, 1996.
19. E. T. Hall, *Beyond Culture,* Anchor Books, New York, 1981.
20. E. T. Hall, *The Dance of Life,* Anchor Books, New York, 1983.
21. D. S. Bosley, Cross-Cultural Collaboration: Whose Culture is it, Anyway? *Technical Communication Quarterly, 2,* pp. 51-62, 1993.
22. E. T. Hall and M. R. Hall, *Understanding Cultural Differences,* Intercultural Press, Yarmouth, Maine, 1990.
23. D. S. Bosley and C. Ross, *Recognition of Cultural Differences Among North American and International Students,* unpublished study, 1995.
24. R. S. Wurman, *Information Anxiety,* Bantam Books, New York, 1989.

CHAPTER 12

The Implications of Translation for Professional Communication

TIMOTHY WEISS

Why should communicators and teachers of professional communication think about translation, a topic that may not touch them at all in their daily work and for which they may have little or no special competency? For some communicators and teachers, the answer to the question is obvious, but for others—probably the majority—translation is an exotic topic out of place among such fundamental business-and-technical communication concerns as audience analysis, effective organization, and clarity. A quick review of the canon of topics of business and technical communication and the current, predominant status of English as an international language of business and the international language of science and technology, lends support to the majority view. As I flip through the dozens of business and technical communication textbooks on my office shelf, I find no index references to "translation"; perhaps a few of these textbooks do mention translation somewhere, but none treats it as a substantial topic meriting explanation, discussion, exercises, and assignments. To my knowledge only one advanced textbook devotes an entire chapter to translation in international context [1]. This absence of translation within the canon of topics of business and technical communication derives from a popular perception that, for speakers of English, an understanding of translation is superfluous: translation is the "other's" concern, not ours.

An understanding of translation, however, is important to professional communicators and to teachers of business and technical communication, and this chapter will argue that the "other's" concern is also "our" concern. Because there will be an increasing volume of information that must be translated from English into other languages and because there will be an increasing number of opportunities abroad for native-speaking English teachers and consultants, professional communicators will find that a basic understanding of translation and a communicative knowledge of a second or third language will indeed be useful—if not now, certainly in the course of their careers. An understanding of the process and difficulties of practical or pragmatic translation inevitably leads to an understanding of the difficulties of information transfer between different cultural contexts; an ability to communicate in a second or third language inevitably opens career doors.

The remainder of this chapter will discuss the prevalence and importance of practical translation around the world and elaborate on two important implications of this translation for professional communicators and teachers of business and technical communication: 1) North American professional communicators can play important support roles in the translation of practical documents in business, government, science, and industry; and 2) North American professional communicators will frequently be responsible for preparing documents that will be translated. The chapter will identify kinds of knowledge and skills that students will need to work in environments in which their writing or documents others have written will be translated into other languages, and it will propose different approaches to helping students acquire these skills. The chapter will suggest directions for research that might promote our understanding of how to meet the needs of working with translators and writing for translation, and it will conclude with a few reflections on the idea of the professional communicator as translator. In short, I will make the case that an understanding of practical translation is important for North American readers; it is a topic that can be addressed in business and technical communication courses and a skill that can be taught in advanced interdisciplinary courses. Interspersed throughout the chapter will be brief commentary drawn from interviews with translators and professors of translation, which I conducted in Rabat, Morocco, in the spring of 1994, thanks to a Senior Fulbright grant.[1]

THE PREVALENCE, IMPORTANCE, AND COST OF TRANSLATION WORLDWIDE

Although scientific and technical literature is largely dominated by English language publication, although telecommunication networks are dominated by

[1] I interviewed ten professors and translators in Morocco who were fluent in English, French, and/or Arabic, and sometimes Spanish. Their responses shed light on the challenges of translation in developing countries and on the potential roles of North American professional communicators to facilitate this translation. For information on the complex language environment of Morocco, see Ennaji [2].

English, and although English is the first or second language of some 666 million speakers, we should make no mistake about the prevalence, importance, and enormous cost of translation worldwide. "Translation goes on at all circuit points in the communication network," Störig emphasizes; "in the international press agencies, in the various departments of newspapers and in radio and television stations; in diplomatic missions and the governments of all countries; in all international conferences; among border and custom authorities; in all export and import firms; in countless military administrative offices . . .; in film studies; in all international means of transportation; . . . just as by all those who . . . warmly receive [the] stream of foreigners in hotels and restaurants, on ships, in factories, at universities, or in the family . . . " [3, p. 19]. In the mid-1980s, the cost of translation worldwide was estimated between 1.57 and 4.73 billion dollars annually, with this figure increasing at the rate of 9 percent per year [4, p. 18]. That puts the total cost of translation today at between 4.8 and 14.5 billion dollars annually. The cost per document or per page will vary, of course, but to give one example of how this huge number breaks down into smaller units, Caterpillar, Inc., whose exports constitute 50 percent of its sales, estimates that translations of its technical literature into over thirty languages cost about $40 per page [1, p. 187]. Another aspect to be considered when thinking about translation costs is the rapidly increasing volume of information material. Document production related to business, government, science, and technology has indeed mushroomed during the past half-century. To draw upon data collected by the American Chemical Society, in 1907 Chemical Abstracts Service (CAS) published 3,437 pages of material; in 1950, it published 8,152 pages; in 1970, it published 48,713; and in 1993, it published 119,000 pages [5]. For countries in which English is not the primary or a widely spoken second language, the cost of translating scientific, technical, and other practical documents can be considerable; moreover, it is cost paid not just by economic powers such as Germany and Japan, but also by developing countries that can least afford to pay the high price.[2]

We begin to get a sense of the pervasive need for translation of practical documents when we consider that although the club of languages in which the majority of scientific and technical documents are written is relatively small, the number of languages spoken in the world is fairly large. "We do not speak one language, nor a half a dozen, nor twenty or thirty. Four or five thousand languages are thought to be in current use," George Steiner remarks in his famous study of translation, *After Babel* [6]. Linguistic anthropologist David Wheeler estimates the actual figure to be closer to six thousand current languages, of which about 10 percent "have enough speakers or enough government support to be considered

[2] In Morocco and Maghreb, for example, a document must often be translated into three languages: Arabic, French, and English. The costs and complications are obvious.

safe from extinction in the coming century" [7, p. A8]. This disappearance of the majority of the world's languages and linguistic richness would leave the number of active languages at 600, which would mean that the task of translation worldwide would still be monumental.

For many societies, translation is not an exception, nor is it a marginal activity; it is a basic, everyday challenge if these societies are to have access to information contained in commercial, industrial, and scientific documents, which today are written primarily in English. We can quickly get a sense of what "primarily" means when we consider a few figures. According to the American Chemical Society, in 1993, 80.3 percent of the abstracted journal literature in chemistry was published in English [5]. That figure represents a steady increase from 43.3 percent (1961), to 58 percent (1972), and to 69 percent (1984). The gap between English and all the other languages is wide, with Russian at the second rank with 6.4 percent. These figures will vary, of course, according to the particular scientific and technical discipline, with some disciplines more and others less dominantly anglophone in language of publication, but the *CAS: Statistical Summary* would seem representative of the dominance of English as the language of publication of scientific and technical literature.

Although these figures might be interpreted as indicating a decreasing need for translation, they equally indicate that everyone who does not read the specialized English of scientific and technical publications will have to rely on translations. The concept of English as an international language has promoted the myth of English as the language that all professionals everywhere speak, read, and write. The reality is different. Wofram Wilss cites a UNESCO study that ascertained that about 50 percent of all publications in the natural sciences remain unread by 50 percent of the experts in the field because they lack the requisite modern language skills. One consequence of this situation, he states, is a duplication of research efforts and a misappropriation of research funds [3, p. 20]. (I recall a comment of a friend, a Uruguayan graduate student in marine biology, who when asking her professor whether she should make the effort to read some of the numerous articles published in Norwegian on her particular research topic, was told to ignore them; since they weren't written in English, their impact would be negligible or non-existent outside of the Scandinavian countries, she was told.) Given that the UNESCO study was undertaken more than two decades ago, its findings would probably have to be revised if it were replicated today. In any case, it would seem safe to say that the notion of English as the universal professional language is an exaggeration; there is a difference between a general capacity to communicate in English (or any other language) and the specialized command of that language needed in order to read materials and write articles in a professional field. It is also safe to posit that, given a choice, most professionals would prefer to read and write a document in their first language. Yes, English is indeed a dominant and prestigious language of professionals around the world, yet the myth of English as international language masks the global reality of linguistic and cultural diversity. It is this reality that North American professionals must prepare for and adjust themselves to—or face the consequences.

EVERY COMMUNICATOR'S NEED FOR A BASIC UNDERSTANDING OF TRANSLATION

In the global workplace and global environment of information transfer and exchange, North American communicators will have increasing opportunities to play important support roles in practical translation;[3] furthermore, even communicators who do not play such support roles will increasingly write documents that will be translated into other languages. It is therefore important for North American professional communicators to understand some rudimentary concepts of translation. These can be simplified into two primary points: first, except for certain universally employed scientific and technical terminology, words in one language do not neatly match up with equivalent words in another or other languages; and second, translation involves interpretation and reformulation of meaning.

In *L'Analyse du discours comme méthode de traduction* (*Discourse Analysis as Translation Method*), Jean Delisle explains that practical translation is not a matching of linguistic equivalencies (words for words, phrases for phrases, sentences for sentences) in two languages; nor is translation the copying of a text from one language into another. Rather, translation is a recursive, heuristic process that involves discourse analysis and interpretation, analogical thinking, and re-formulation and re-expression of the ideas of the source text. The source and target texts are therefore not equivalent in any linguistic sense; they are, rather, more or less analogous at a basic semantic level of ideas or information content [10, p. 85]. Similarly, in *Fondement didactique de la traduction technique* (*A Didactic Foundation for Technical Translation*), Christine Durieux emphasizes that the technical translator does not translate "a succession of words," but rather, a message that first of all must be understood—analyzed and interpreted—before it can be reformulated and re-cast into the respective language and culture [11, p. 27]. Delisle calls practical translation an intellectual operation that consists in reproducing the articulation of a thought in a discourse; the art of translation is based on a heuristic ability to observe "the non-parallel structures of two languages" [10, pp. 44-45].[4]

[3] Practical or pragmatic translation constitutes a type or category. In *Translating Meaning*, Eugene A. Nida [8] lists four types of "interlingual communication": business, political, technical, and literary. For more information on practical and pragmatic translation, see [9].

[4] Nida believes that this ability depends on a certain range of language learning: "Potential translators . . . should know at least three languages. If such persons know only two languages, namely a mother tongue and one foreign language, they are likely to become rather mechanical in setting up types of correspondences" [8, p. 91].

My Moroccan interviewees concurred that translation often necessitates the transformation of a text; the translator must often make decisions about what to include and exclude, what to expand and what to summarize. Anytime there is a difference of knowledge or culture between the source discourse community and the target discourse community, translators will respond to this difference by transforming the text accordingly. In correspondence, for example, translators may decide that it is unimportant to retain certain formulas of courtesy and politeness that occur in one language but not in another (e.g., these formulas are common in business letters in French and Spanish, for example, but not in English). Conversely, in scientific and technical documents, differences of knowledge and culture between source and target discourse communities may create a need to explain concepts or cultural particularities to the readers, or they may bring to light an element of the text that can be deleted without reducing its essential meaning.

Translators must know languages and their respective cultures well enough to recognize non-parallelisms of meaning and structures; they must also have the ability to re-express ideas in the target language, which will usually be the translator's native language. This latter qualification is not one that can be assumed or taken for granted: the "majority of the English mother-tongue applicants for translation posts in the European Commission fail *because of the poor quality of their English*" [12, p. 18]. One, but not the only, challenge of translation, then, is re-formulation and re-expression in the translator's native language: translators must be good writers; this is true no matter the language, and doubly so the more the source and target languages and their cultures differ. In *Thinking Translation: A Course in Translation Method: French-English,* Hervey and Higgins begin with exercises that test students' ability to reformulate and recast ideas in their first language [12]. As I will explain in the next section of this chapter, this is a promising exercise for our students as well, for it can teach a fundamental concept not only of translation but also of business and technical communication: adaptation of a text to a designated audience.[5]

NORTH AMERICAN COMMUNICATORS' ROLES IN THE ENTERPRISE OF WORLDWIDE TRANSLATION AND THE KNOWLEDGE AND SKILLS NEEDED TO FULFILL THOSE ROLES

Depending on their content-area knowledge, language skills, and cultural expertise, North American communicators will play a variety of roles in the enterprise of worldwide translation. Some will themselves be interlingual

[5] The trend in translation theory is to view translation as an activity across cultures [13-15]. Nancy L. Hoft [1] emphasizes the collaborative nature of translation projects. Practical translation often calls for a specialized content-knowledge, although it is important to note, as does Durieux [11], that numerous are the texts that appear to treat one content area but in fact also treat another or others.

translators, others will facilitate the operation of translation by functioning as consultants and coordinators, while others will continue to produce documents that will eventually be translated into other languages. Finally, all professional communicators who go abroad—or who work with international colleagues—will be forced, by the contexts in which they will inevitably find themselves, to translate, to communicate as best they can in other languages.

Given that the majority of practical documents that are translated throughout the world are translated from English into other languages, the majority of tomorrow's translators will not be North American. What will be the roles, then, of North Americans in the realm of international translation? Because practical translation often necessitates collaboration and team projects, even if an individual is not sufficiently qualified to translate documents, there are other tasks and roles within the team for those with appropriate language and managerial skills. In *International Technical Communication: How to Export Information about High Technology,* Nancy L. Hoft [1] discusses the various roles of the translator, the technical communicator, the translation coordinator, and the translation team leader, for example. The list could be easily expanded, but it could also be simplified into two groups: the interlingual translators themselves and those who coordinate and consult on the translation project and its products. While some North American communicators will themselves become professional translators, especially those with expertise in Spanish, French, or Portuguese (along with English, the languages of the Americas with the largest number of speakers), the majority will serve as project coordinators and consultants, for example, collaborating with translators whose native language is that of the target audience, be it Chinese, Japanese, or Arabic. They will help interpret problematic semantic, rhetorical, and cultural aspects of the texts, and they will "translate" documents in American English intralingually into an "international English" and redesign them so that the interlingual translators can do their work with greater accuracy.

Beyond this specific, formal, professional domain of translation of practical materials there still remain all those other contexts, as Störig notes, where translation takes place informally. To work and live in a society whose language and culture are different from one's own is to be called to translate every sign every moment every day. North American professionals who work in other countries will need to be translators in this latter sense, and in order to be successful they will need to understand the process of translation, whether that process is applied to a text, an informal conversation, or non-verbal signs. Even if their close professional colleagues all speak English, they will likely need to interact with other staff members who do not, or risk being isolated and left out of activities. In international contexts, translation is not just the business of specialized translators; it becomes, by the nature of the context, everyone's business.

What kinds of documents are translated worldwide, and what knowledge and skills do North American professional communicators need to play important roles in the translation enterprise? To answer these questions I will draw in part on information gathered from interviews conducted in Morocco, a country in which practical translation is an everyday concern and challenge. The kinds of documents translated include conference abstracts, scientific and technical articles, correspondence (governmental and non-governmental), speeches, newspaper releases and features. The clients included banks, the mass media (newspapers, radio, television), governmental and non-governmental institutes of various sorts, the national government, embassies, and world organizations such as the Arab League and the United Nations. According to one interviewee, who spoke from the viewpoint of a university educator and administrator, the greatest needs for translation in a developing country like Morocco pertain to two categories of document: scientific articles and technical documents that bring new knowledge and its applications, and basic textbooks on science and technology for primary, secondary, and university students. Throughout the world there are many countries like Morocco in which English is not a first or widely spoken second language; in these countries the volume of potential information transfer in business, government, science, and industry, creates new opportunities for North American professional communicators and new demands of knowledge and skills.

To take advantage of these new opportunities and to respond to the new demands of information transfer in a global context, today's professional communicators need to develop six categories of knowledge and skills: superior English language skills; general knowledge linked with specialized content-area knowledge; research skills; analytic, interpretive, and rhetorical skills; computer skills; language-learning and other cross-cultural skills.

Source language skills are indeed the *sine qua non* of professional communicators; communicators facilitating translators—or writing documents that will be translated into other languages—must themselves write well and have a good command of English. They must know how to express concepts and ideas in various ways depending on the context in which a document will be used and on its users' different levels of knowledge. As Hervey and Higgins note, these skills are not to be taken for granted [12, p. 18]; similarly, Moroccan interviewees concurred that poorly written source texts were the number-one problem confronting translators. If a source text's ideas are unclear, or if the text is badly organized, understanding it turns into guess work, not translation, and recasting and re-expressing the text will likely deviate from the intended meanings. It is hardly the translator's job to be the author's editor. But in these cases, professional communicators who are native speakers can serve as editors, explaining to translators who are not native speakers of English (or who do not possess the specialized content-area or cultural knowledge) what the author probably meant by this or that expression and how the text could be better presented.

Depending on the context and project, other kinds of knowledge and skills may be just as, or even more, important to the professional communicator who coordinates or consults on translation work. Here are some observations and considerations pertaining to this knowledge and these special skills:

- A wide general knowledge of North American culture and specialized content-area knowledge enables communicators to help translators overcome difficulties with specialist terminology, idioms, and other culturally related elements (e.g., a reference to "class rings" as one of the products of an American company). Although translators in other countries can sometimes consult computerized databases of scientific and technical terms, a further explanation of the term may still be required.
- Because knowledge is increasingly interdisciplinary, even if communicators work with documents of a single content area, they will still need the skills to research other related areas that will bear on the translation of these documents. This kind of background research demands many different informational skills; it demands an ability to discriminate between essential and non-essential reading material for a particular translation project and, in general, it demands a curiosity about the world.
- Communicators must have the analytic, interpretive, rhetorical ability to comprehend the pattern of ideas within a text as "extra-linguistic" entities for which the language constitutes but the vehicle. They must possess a basic understanding of translation and the ability to rewrite and edit a text as a collection of ideas rather than as a collection of words.
- Communicators will increasingly make use of the gamut of computer skills related to machine-assisted translation (MAT).[6]
- Language-learning skills are obviously important; however, contrary to public belief, not even translators—much less those who consult with translators or coordinate translation projects—need be perfectly bilingual. To translate practical documents one must have a good passive command of the source language and an excellent command of the target language. To consult or collaborate with translators one may not need a knowledge of the target language, but a knowledge of the target language would certainly be a plus, if only in terms of the professional relationship that the consultant or coordinator has with the translators. This is partly a matter of mutual respect for one another's professional capabilities and different linguistic, cultural backgrounds. North American communicators who work and live abroad will need language-learning and cultural skills in order to participate in the social element of the professional environment and to function within the society.

It is worth noting the responses of Moroccan interviewees to questions about the knowledge and skills of prospective translators because they also shed light on

[6] Writing in the early 1990s, Patricia Thomas states that the translator's workstation will include: "a word processor with multi-window screen from which one or more term banks may be consulted and access given to an MT system, with the possibility of interactive prompting either on screen or via speech synthesis: desk-top publishing for the final product; and the ability to receive and transmit text through electronic mail networks or facsimile transmission" [16, p. 144].

qualifications of those who collaborate and consult with translators. Although knowledge of the source and target languages is, of course, an absolute prerequisite qualification of translators, interviewees also noted that the more content-area knowledge one has, plus the more experience one has living in source and target cultures, the more adept one can become at resolving certain difficulties of translation. Intelligence and resourcefulness were emphasized as important qualifications. Translation demands not only the analytic and interpretive ability required to understand the pattern of ideas of a source text, but also the ability to re-express those ideas analogously within the world view of the target audience. Translators must perceive differences as well as similarities; they must shuttle between the two worlds that they have brought together by the analysis and interpretation of the source text and the re-expression of its ideas in the target text. A translator must be a deft and swift researcher, knowing where to look and whom to ask when the answer to a particular problem is beyond one's personal cognizance. One interviewee called translation a "gift"; what he meant, I believe, is that translation is a demanding process, creative as well as analytical; it involves much more than flipping through specialist dictionaries and substituting a word in one language for an "equivalent" word in another, for finally there are no equivalents, except for certain scientific and technical terminology that has gained a universal usage.[7]

TEACHING APPROACHES AND RESEARCH TOPICS

What approaches can teachers of business and technical writing take to bring the implications of worldwide translation into the classroom, and how can the rhetoric and professional writing curriculum be developed to address the knowledge and skills that students interested in pursuing careers in international communication and/or practical translation will need? Here are some considerations and partial answers to these questions.

Interpretation, Analysis, and Transformation of Texts

Too often our approach to teaching business and technical communication focuses on the words, rather than the ideas, of a text. To teach the implications of translation it would be better to begin with finished documents and then to break down those texts, through analysis and interpretation of their content and form, into an outline of ideas; once the outline is established, students could then be directed to reformulate and rewrite the text for an audience that has a different

[7] Although a few interviewees were of the opinion that the translator of scientific and technical documents does not need a knowledge of the target culture, other interviewees vigorously disagreed. They insisted that, even for a translator of scientific and technical documents, "real contact [with the target culture] is important."

knowledge base and cultural, social background than the audience of the original document. This kind of exercise can be brief or extensive, the latter involving researching a particular culture or geographic region and interviewing international students from that culture or region.

Some variations on this exercise might include "translating" texts from idiomatic American English to internationally standard English, translating a text from British into American English, and rewriting for high-context audiences texts originally written for low-context audiences [17].

An Audience Analysis Unit Incorporating International Audiences

Before suggesting ways to incorporate international audiences into professional communication courses, I would like to summarize some comments that my Moroccan interviewees made about the centrality of audience to translation: they concurred that audience is the primary consideration in the translation of practical documents; translation would be a perpetual somersault without the grounding reality of audience to guide the translator in choices at all different levels. Interviewees talked about the choice that translators must often make between simplifying the message, and allowing it to retain more of its nuances for content-knowledge audiences. Translators must strike a balance between faithfulness and accessibility. Discussing how the audience's cultural expectations can shape translation, one interviewee cited an airline advertisement whose theme was punctuality; since, in many Arabic cultures, hospitality, not punctuality, is what matters most, if the airline wanted to effectively translate the ad for an Arabic target culture, it would probably want to alter the theme of the ad to better address the target audience's value system.

The gist of the Moroccan interviewees' comments support our dominant approach to teaching business and technical communication, which focuses on audience as the key concept. What we must do now is extend the concept to include multicultural and international audiences. I have already suggested exercises that treat the differences between American English and an internationally standard English; editing documents to eliminate "bureaucratese" is another related dimension of internationalizing a text that applies particularly to business correspondence and reports. Almost any exercise or assignment can be defined as one in which the audience of the document is potentially international.

One approach to teaching translation is to pair a North American student with an international student. For example, the North American and the international student each would write an ad for the same product, one ad directed toward a North American audience and the other toward an audience in the international student's country. Once the ads have been written, the students could explain their strategies and the factors that determined their construction of the ad in a particular way. Working with another product or service, the students would then change roles, with the North American student writing an ad for the international audience and the international student writing an ad for a North American audience.

As I have noted, our business and technical communication textbooks lack a conceptual framework in which to present a discussion of the issues of translation and international communication. Some textbooks ignore the topic altogether, as if it is tangential to what communication is supposed to be about, while other textbooks treat it cursorily. We need a framework that extends the key concept of audience to international contexts and inquires into the universality or non-universality of the rhetoric and document design of North American pragmatic documents. We need a framework that provides us with a method of analyzing the interaction of local and global culture in varying contexts and the consequent difficulties of translating discourses that project into different cultural contexts. Perhaps most of all we need a framework that enables us to address the individual in international contexts, rather than the categorical Chinese, Japanese, or Arabic audience. Generalizations about cultures are good to know, but finally it is the individual who responds to communications and that individual, whose language and culture will often be different from our own, will be just as complex and just as unpredictable as any individual in our own society.

An Expanded Role for International Students and Satellite Television

One means of bringing an international component into the business and technical communication curriculum is to draw upon a sometimes marginalized human resource: international students in our business and technical communication courses and at our colleges and universities. Along with satellite television, which makes it possible to enter a virtual reality of another language and society thousands of miles from one's home, international students bring the world to our campuses and help create a multilingual, multicultural learning and living environment. In 1982 there were 331,000 international students enrolled at U.S. colleges and universities; by 1992 the figure had risen to 458,000. At the graduate level, 11 percent of all students at American colleges and universities were international; in science and engineering programs, 43 percent of the graduate students were international [18; 19, p. A33]. This is an impressive figure, and teachers of business and technical communication have only begun to recognize what this means to the field and how to shape courses and text materials accordingly [20, 21].[8] We need to draw more actively upon these students' linguistic and cultural diversity in framing exercises, assignments, and case

[8] Teachers of professional and technical communication who are interested in intercultural and international issues in communication can benefit from membership in TESOL (Teachers of English to Speakers of Other Languages) and in participation in its ESP group (English for Specific Purposes).

studies of an intercultural and international scope; as I have suggested, international students are resources to draw from in our business and technical communication courses, and these students can play roles as assistants in translation courses and as adjunct teachers in courses of important but less-commonly taught languages at American universities, such as Chinese, Arabic, and Japanese.[9]

Teaching Practical Translation in Special Interdisciplinary Courses

There are some basic changes in curriculum and orientation that can be taken to bridge the gap between the skills and knowledge of today's business and technical communication students and those that they will need in order to play significant roles as communicators in international contexts. Fortunately, in order to teach practical translation, colleges and universities do not have to begin at zero. Many, if not most, students at American colleges and universities must satisfy a minimal second-language requirement, and many, if not most, colleges and universities offer intermediate-level courses in business Spanish, business French, and similar courses in other languages. In these courses, or in a course at a higher level, practical translation could be taught, focusing on those kinds of communications in which the human interaction is most important (ads, letters, memos, telephone conversations, short oral presentations, and so forth). At an advanced level, translation of specialized documents within the student's discipline could be treated; in such an interdisciplinary course, a language specialist, international students working as assistants to the language specialist, a professional communication specialist, and a content-area specialist might all play a part.

It is time to be frank about who we are and what we must learn in order to compete in an international marketplace. In the future, we will have to learn other languages in order to remain competitive. It can be done. The junior-senior level students whom I taught in North Africa knew three and sometimes four and five languages; similarly, European students undergo much more rigorous training in languages than do American students. American society, in conjunction with its colleges and universities, needs to make language instruction a top priority.

Research Topics

In the comments above I have already suggested some directions for research about translation and internationalization. For all kinds of documents, researchers

[9] French linguist Claude Hagège has recommended that the French school system turn its attention to teaching these, as well as other non-European languages, not only because Asian nations have become economically powerful, but also because learning about non-European languages and cultures breaks down xenophobic sentiment [22].

need to inquire into the cultural differences of audience expectations. For example, North Americans have certain expectations when they begin to read instructions, but do audiences of North Africa or other regions of the world share them? How do readers around the world read practical texts? Are there some rhetorical universals by which writers and translators can be guided? Is multilingual advertising and product documentation a necessity, and is it the best solution to communication within a zone of linguistic plurality?[10] How do other countries respond to the necessity of translation, and what priorities have they set?

FINAL REMARKS

Every word is a translation; every communication is a transformation. Professional communicators have always been translators in the basic sense that whenever the context of an idea is changed, the idea itself must be translated in order for it to be understood and effectively presented. The professional communicator bridges, translates ideas between, different contexts. This broad sense of the meaning of translation enables us to extend our current notion of the identity of the professional communicator to encompass the new necessity in a borderless world for communicators to shuttle between linguistically and culturally different contexts. Translation can serve as a metaphor for changes that are reshaping business and technical communication issues and concerns. The ultimate implication of translation for professional communicators, then, is that it transforms their very sense of who they are and who they must become in the future.

Translation is not an exotic flower, for its implications touch basic concepts, such as audience and discourse analysis, of our business and technical communication courses. With a focus on translation, a teacher and researcher can address any issue in professional communication. I hope that teachers' and communicators' new awareness of the importance of translation worldwide and its expanding study[11] will finally help us accomplish two things: first, to change our courses from ones that focus much too narrowly on prescriptive procedures and principles to courses that emphasize analysis, interpretation, and transformation of messages; second, to reinvigorate the study of modern languages within the rhetoric and professional communication curriculum as well as within the university curriculum in general.

[10]One simple yet complex exercise might involve translating labels for products into bilingual or multilingual labels; this in turn could lead to research on the efficacy or inefficacy of multilingual labeling and product documentation.

[11]Here in Hong Kong, for example, six tertiary institutions now teach specialized courses in translation.

REFERENCES

1. N. L. Hoft, *International Technical Communication: How to Export Information About High Technology,* John Wiley and Sons, Inc., New York, 1995.
2. M. Ennaji, Aspects of Multilingualism in the Maghreb, *International Journal of Sociology of Language, 87,* pp. 7-25, 1991.
3. W. Wilss, *The Science of Translation: Problems and Methods,* Gunter Narr Verlag, Tübingen, 1982.
4. N. Richert, *Arabisation et Technologie,* L'Institut d'Etudes et de Recherches pour l'Arabisation, Rabat, 1987.
5. American Chemical Society, *CAS: Statistical Summary, 1907-1993,* American Chemical Society, Columbus, Ohio, 1994.
6. G. Steiner, *After Babel: Aspects of Language and Translation,* Oxford University Press, London, 1975.
7. D. L. Wheeler, The Death of Languages, *The Chronicle of Higher Education, 16,* pp. A8-9, April 20, 1994.
8. E. A. Nida, *Translating Meaning,* English Language Institute, San Dimas, California, 1982.
9. J. Newton (ed.), *Computers in Translation: A Practical Appraisal,* Routledge, London, 1992.
10. J. Delisle, *L'Analyse du Discours Comme Méthode de Traduction,* University of Ottawa Press, Ottawa, Canada, 1980.
11. C. Durieux, *Fondement Didactique de la Traduction Technique,* Didier Érudition, Paris, 1988.
12. S. Hervey and I. Higgins, *Thinking Translation: A Course in Translation Method: French-English,* Routledge, London, 1992.
13. M. Snell-Hornby, *Translation Studies: An Integrated Approach,* John Benjamins, Amsterdam, 1988.
14. E. Gentzler, *Contemporary Translation Theories,* Routledge, London, 1993.
15. A. Lefevre, *Translating Literature: Practice and Theory in a Comparative Literature Context,* Modern Language Association of America, New York, 1992.
16. P. Thomas, Computerized Term Banks and Translation, in *Computer in Translation: A Practical Appraisal,* J. Newton (ed.), pp. 131-146, Routledge, London, 1992.
17. E. T. Hall, *Understanding Cultural Differences,* Intercultural Press, Yarmouth, Maine, 1990.
18. *Almanac of The Chronicle of Higher Education, 41*:1, 1994.
19. A. M. Rubin, Foreign Influx in Science Found to Cut Americans Participation, *The Chronicle of Higher Education, A33,* July 14, 1995.
20. T. N. Huckin and L. A. Olsen, *Technical Writing and Professional Communication for Nonnative Speakers of English* (2nd Edition), McGraw-Hill, New York, 1991.
21. L. Trimble, *English for Science and Technology,* Cambridge University Press, Cambridge, 1992.
22. C. Hagège, Parlez-vouz Européen? An interview with Dominique Simonnet, *L'Express, 2161,* pp. 48-51, December 11, 1992.

CHAPTER 13

The Student-As-Researcher in International Organizations

JANE JORGENSON

To understand the communication processes and practices that are unique to the intercultural workplace, there is a need to consider how its members make sense of their environments—how people working in particular intercultural settings come to understand and manage the activities and occasional contradictions of their day-to-day experience. Such learning about the international organization "from the inside out" [1], or from what Malinowski long ago referred to as "the native's point of view" [2], is a difficult enterprise. It typically involves the direct and prolonged engagement of the researcher in the organization by means of observation, conversation, interviewing, and other field methods.

For those organizational researchers who seek to represent and interpret members' perspectives qualitatively, the research process raises several recurring questions: for example, how does one identify significant patterns in qualitative material? how does one construct adequate descriptions of organizational life? what is the relationship of the researcher to the researched? and how is what we come to know in organizational fieldwork affected by who we are, including our personal characteristics and past experience? The latter two questions, in particular, pertain to point-of-view issues and research relationships; they invite us to consider our roles as researchers, to see ourselves as participants in the communicative construction of our findings, rather than merely as detached observers, and to reflect on our social and emotional involvement in the setting as part of the data that we analyze [3, 4].

INSIDER/OUTSIDER TENSIONS IN ORGANIZATIONAL RESEARCH

The nature of personal involvement between the organizational fieldworker and research subjects has long been a focus for discussion and controversy. Part of this tension derives from the requirements placed on the fieldworker to operate as both "insider" and "outsider" to the organizational setting; to cultivate familiarity and rapport with the research participants while avoiding the temptations of overrapport (also known as "going native"). In acquiring progressive familiarity with the setting, the fieldworker as insider tries to apprehend members' understandings of the situation they are in, while as an outsider the task is to articulate (to other scholarly outsiders) what one's informants take for granted but have no need to say [1].

Some authors suggest that the study of corporate cultures is hampered rather than helped by similarity and familiarity between the fieldworker and organization members—similarity of culture, group membership, or other frames of reference. In a recent essay, McLeod and Wilson, for example, allude to the long history within anthropology of the *non*native observer conducting fieldwork in non-Western cultures. Within this tradition, "familiarity" and "similarity" between observer and observed are problems which potentially limit any valid description in corporate culture studies. Therefore, McLeod and Wilson argue,

> the obligation of the anthropologist and the student of corporate culture alike must still be to distance their activities as analysts clearly from the objects of study [5, p. 286].

Such arguments are consistent with the perspectives offered in many research methodology texts and courses where the organizational researcher is advised to adopt the stance of a neutral observer who strives to maintain analytical distance from his or her research subjects.

While such a conception of researcher roles assumes that cultural distance ensures more valid description, it fails to acknowledge the socially situated and transactional nature of all research activities; as the researcher tries to make sense of informants' accounts and explanations of everyday activities in the organization, informants are similarly engaged in making sense of the researcher and the purposes of the proposed research.

THREE RESEARCH ENCOUNTERS

In this chapter I draw on the experiences of individuals engaged in field research in international organizations in order to consider some of the rewards and difficulties for the researcher who is both an "insider" and "outsider" to the research setting. Unlike the nonnative observer in exotic settings, the student of complex organizations functions as a kind of "endogenous researcher" [6], bringing to the inquiry an "insider's" knowledge. Such knowledge can derive from many sources including nationality, cultural and linguistic familiarity, and work experience in organizations similar to the research setting.

What is distinctive about the experiences of the individuals described in this chapter is that each is a foreign national studying in the United States who chose to return "home" to conduct organizational research. In the case of one student, this involved interviews with Chinese nationals living in the United States and a return to China to conduct participant observation in a large manufacturing organization. For a German student, it meant gaining access to German-owned companies in the United States to carry out interviews on cultural differences as perceived by German and American employees. A third case, that of a Finnish national living and working in the United States, also involved a combination of travel back to Finland to carry out interviews in a Finnish organization, and parallel research conducted in the United States. Each student researcher, after living in another culture for a period of time and encountering new symbols and new definitions for familiar symbols, has had to confront the old and predictable from a new vantage point. In this chapter, I retell parts of their stories in order to problematize the research process as it develops from literature review, to entry into the research site, to analysis of the fieldwork experience. Despite differences in cultural context and research focus, each story illustrates how the researcher's group memberships and interpersonal transactions with informants profoundly shape the research process and eventual findings. My aims in recounting their experiences are: 1) to examine some of the unique advantages and complications that derive from a researcher's identity as a cultural "insider," 2) to articulate, more generally, the benefits to the researcher of adopting a reflexive stance, in which one reflects on one's starting points in order to make explicit some of the frameworks of meaning that shape the research process and eventual findings, and 3) to consider how a reflexive mode of inquiry contributes to a fuller understanding of the specialized contexts of international professional communication.

FORMULATING THE PROBLEM: PRIOR TEXTS AS STARTING POINTS

A significant problem facing researchers in cross-cultural settings is that their understanding of intercultural contexts is often shaped long before they enter the field by their reading of relevant literature. Using prior texts as starting points for research can present a quandary for the field researcher on several counts. On one hand, literature reviews can help to focus a problem, "to articulate compelling and researchable questions" that are salient to a professional audience [7].

More particularly, the reading of prior texts aids the researcher in cultural description and analysis when it provides a "cross-cultural metalanguage" [8], a vocabulary of higher-order abstractions or dimensions along which cultures are likely to vary. Hall and Hall's [9] concepts of monochronic and polychronic time, and of high-context/low-context communication are well-known examples of a metalanguage which serves to "sensitize perception," leading the observer "to look for things [or *kinds* of things] that might be important in a particular situation" [8, p. 223].

In presenting us with sensitizing concepts, the texts that we read before fieldwork structure the possibilities of our experience, thus shaping the analysis and our subsequent findings. More simply, the literature determines which categories we are willing to see, and this both constrains and facilitates our inquiry. The dilemma for the researcher is that texts read in advance can reduce openness to what emerges in the field. When those texts present broad-based cultural classification systems, they can lead the observer to overgeneralize.

In comparative studies there is often a temptation to treat cultures as if they were homogeneous. This tendency to overstate the degree of "sameness" in a group in order to enhance between-group contrasts unfortunately serves to obscure what could be fundamental differences and tensions among individual group members. In those cross-cultural studies which borrow the approaches of quantitative social research, analysis usually proceeds by aggregating the responses of many subjects in order to arrive at group generalizations; the result is a merging of the identities of many into a homogeneous "other" which can then be collectively labeled as "Japanese managers" or "French factory workers" [10, p. 802]. One such framework which has been especially influential in organization studies is Geert Hofstede's widely read work on core cultural values [11, 12]. Based on survey responses from IBM employees in fifty-three countries, Hofstede's findings present the relative positioning of various countries on four value dimensions: power distance, collectivism-individualism, femininity-masculinity, and uncertainty avoidance.

In the scope of its database and the simplicity of its framework, Hofstede's work has provided managers and other practitioners with an important first orientation to issues of cultural difference [13, 14]. Based as it is, however, on assumptions about the collective identities of national groups, work such as Hofstede's tends to support a monolithic view of culture. As Sackmann points out, any member of an organization is a member of multiple societal subcultures based on such features as ethnicity, religion, profession, and gender [15]. These various roles give rise to different involvements and concerns depending on the tasks, issues, and problems at hand.

Some critics have also pointed to the errors in comparative analyses organized around binary oppositions—as when Japanese collectivism is presented as the "opposite" of American individualism. Hamada questions the logic underlying such dimensions given that "these phenomena are each made up of diverse components, internal counterpoints, and variations" [16, p. 199]. Similarly, Bateson argues that our tendency to view social differentiation in terms of simple bipolarities blinds us to the presence of other, more complex patterns and motifs [17]. Criticisms such as these underscore the dangers of theoretical abstractions and serve as a reminder to use the concept of culture "gently." According to Pearce, "with a sufficiently detailed knowledge, it is possible to deconstruct and deny any sharp distinction that someone might draw between cultures" [18, p. 50].

ANOTHER STARTING POINT: INTERPERSONAL RELATIONSHIPS IN FIELDWORK

Survey-based research such as Hofstede's is marked by a concern with generalization: that is, with establishing the relevance of findings beyond the boundaries of a particular investigation. From this point of view, the researcher who administers surveys is advised to present a uniform "self" across encounters with many respondents. In this way, detachment and neutrality help to ensure reliability of results by reducing the possibility of bias or reactivity.

Qualitative field studies, in contrast, focus attention on the "diverse forms and details of social life" [7, p. 21]. A key characteristic of field research is that it uses human investigators, rather than standardized instruments, as the primary vehicle for data gathering. Participating in relationships with the members in the setting is the basis of the interpretive process which is so central to field research. Thus relational issues—the developing interpersonal relationships between the observer and the observed—provide an important point of departure for any assessment of the quality of data collected, as well as for its interpretation.

Problems at Entry as Data: Teja

Fieldwork relationships are usually initiated when the researcher enters the organization. Crossing organizational boundaries in order to observe day-to-day or ceremonial activities or to conduct interviews often requires substantial negotiation with organization "gatekeepers," and the difficulties of gaining acceptance are a common preoccupation.

Although we tend to think of entry as merely an antecedent step to fieldwork, entry experiences can be an important contextual resource in the interpretation of subsequent material, as illustrated in Sutton's research on dying organizations. He initially contacted twenty companies but was refused access by twelve. His published account includes his reflections on the repeated refusals, attempting to make sense of their implications for later data gathering and analysis [19]. In some cases, the resistance of members to researcher access can lead to rethinking what should "count" as data. In her research on women artists, Kauffman kept notes on her negotiations for interviews with prospective subjects, material which then became useful as "a map *for* the analysis rather than a qualifier of it" [20, p. 187]. She eventually came to interpret the refusals and distancing moves of some of the artists in terms of the larger themes of the study: that is, the ways in which artists symbolically protect the boundaries between "woman" and "professional."

The three students who sought cultural "reentry" in international organizations encountered wariness and occasionally skepticism in their initial contacts. Entry problems were especially acute in the case of the German student, Teja, who attempted to gain access to German-owned and -managed companies in the United States [21]. Teja's study was intended as an exploration of decision-making and planning processes involving German managers and American employees. His idea was:

> to get people in organizations to reflect on a specific situation in which Americans and Germans had been acting together to make a decision. This reflective process was planned to be focused on cultural issues that in any form, hidden, or ópenly, had affected the interaction [21, p. 40].

He contacted the German managers of four American subsidiaries and arranged personal meetings for the presentation of his research proposal. However, in each of these meetings, after being introduced to the research idea, the German managers responded by denying any significant impact of cultural differences in their organizations. Each of the four concluded that since relevant cultural differences between Americans and Germans did not appear to exist, the proposed study would be a waste of corporate time. As one manager said, unless the researcher was interested in "more meaningful areas like differences in the legal or financial environments," he did not feel that the company was in a position to support the research [21, p. 46].

This initial resistance led Teja to redesign the project to focus on the individual manager, rather than the organization, as the unit of analysis. He subsequently located a group of managers in five other companies who agreed to share their experiences and personal reactions "as Germans on assignment in America."

The resistance of the four German managers makes clear that in cross-cultural research the indigenous researcher is not guaranteed an advantage. In this case, shared nationality was not deemed by the managers a sufficient basis for the establishment of trust and rapport. Indeed, one of the managers commented after their initial introductions, " 'Teja—that's not a German name, is it?" thus, from Teja's perspective, using the issue of names to challenge indirectly his authenticity as a German [22].

After some reflection, Teja constructed various interpretations of the initial refusals. All of the managers had been living in the United States for at least twelve years; had they perhaps adapted their behavior to their daily environment to an extent that they had lost an awareness of cultural contrasts? Or did their reluctance lie in their view of him—did they question his naivete and lack of experience with real problems? Did the research proposal evoke anxiety because they saw the researcher's role as ambiguous and potentially disruptive? Did they fear that the proposed study could uncover problems and therefore disturb the smooth running of the organization?

Beyond his difficulties in establishing trust, Teja's experience illustrates several points concerning the process of interpretive research. First, it is essential to understand how the research subjects or co-participants make sense of the researcher's aims. In this case, the topic of cultural difference may have been viewed with skepticism by organization members because of its abstractness and because of the absence of a convenient language (or metalanguage) in those organizations for discussion of the issues. Second, the topic, in its presentation, would have been perceived as open-ended, focusing on the subjective experiences of individuals. In this sense, the results of the research were potentially problematic because they would be seen as outside the organization's control.

The manager's resistance could also be seen as an expression of the "managerial metamyth" [23], a bias toward rational-technical values which is characteristic of modern work organizations. One of the assumptions of the managerial metamyth is that all work processes can be fully analyzed and controlled. Cultural issues cannot be easily dealt with within this framework, because they are not readily "programmable."

The Impact of Group Memberships: Liisa

Some organizational researchers have argued for more systematic study of how personal factors enter into the research process; how, for example, does our past experience guide our topic choice, and what perspectives are lost or gained by virtue of our personal orientation to those topics. Not only do personal interests incline us initially toward certain problems and questions, but, as Peshkin suggests, they often lead us to "take sides" [24].

Liisa's experiences as a Finn engaged in fieldwork in American and Finnish banks illustrate how personal attributes can serve as reference points in research [25]. Her dissertation project focused on responses to change and uncertainty in the workplace as experienced by American and Finnish employees. In explaining her topic choice, she emphasizes two features: her bicultural, bilingual upbringing, having attended American and Finnish schools as a child in Finland, and her years of employment in banks, first in Finland, and then in the United States. She recognizes that her language skills and work history inclined her to choose a research context in which she could bring to bear her unique expertise.

Another potent influence on her research approach and point of view is her membership in one of the organizations to which she sought access for the research. At the time of the study she worked as an associate level (non-supervisory) employee at an American bank—an environment she often described as "pink collar" in reference to the gender distribution of its employees. Liisa views her dissertation focus as a response to an overemphasis on managerial viewpoints in the organizational change literature, from which, she says "the little person was missing" [26]. This accounts for her decision to include the perspectives of workers as well as managers in her interviews.

Although Liisa did not work in the same division as her American research respondents, she was intimately familiar with their work environment. On reflection, she perceived the effects of this proximity as contradictory. In some cases, she felt it created a basis for increased rapport and greater frankness on the part of respondents. At the end of the interviews, after the tape recorder had been turned off, two respondents expressed their enjoyment "at having been able to 'unload' freely to someone who understood what they were saying because she worked with the company" [25, p. 153]. This reaction was not uniformly shared. Even though Liisa stressed to prospective participants the confidentiality of all interview material, her status as employee raised the concerns of another interviewee who feared that Liisa might divulge personal comments to management.

A formal organizational position such as Liisa's provides the researcher with an unusual vantage point for investigating the local culture of the organization. Her membership, *as perceived by organization members,* influences the quality and depth of their participation through their willingness to respond to her questions. Like Teja's problems at entry, Liisa's various encounters with willing and reluctant participants can be an important source of data insofar as they indicate the sensitivity of specific issues and the ways that intra-organizational boundaries (for example, the boundary between "management"/"non-management") are perceived.

Redefining the Researcher's Identity: Tang

Textbooks on research methodology tell us that the most productive research relationships are those characterized by trust and rapport—these qualities are seen as essential for eliciting "candid" and full disclosures of information from the research participants. To obtain the cooperation of respondents, the researcher must skillfully manage the research encounter; to the extent that she appears "supportive, cordial, interested, nonargumentative, courteous, understanding," she is likely to obtain the fullest cooperation of research participants [27].

Yet methodology texts tend to overlook that the meaning and establishment of trust is culturally situated. There are, for example, settings in which the interviewer's individual self-presentation is less important than the discovery of shared social connections or past associations. The Chinese term "kuan-hsi," meaning "social relations," refers to the use of interpersonal networks to build intimate relations with comparative strangers [28]. Americans evoke a similar idea when they speak of finding common ground—as when two women find they are alumnae of the same college. In the Chinese context, however, kuan-hsi extends the notion of "personal contacts" to include a stronger element of instrumentality such that these social resources (and their implied obligations) can be used to achieve personal and professional goals.

Tang's research explored the consequences of management reform in Chinese businesses, focusing on the transferability of American organizational practices to Chinese settings [29]. His father, a prominent businessman, used personal contacts to facilitate his son's entry into a large industrial organization where Tang was allowed to carry out observations and in-depth interviews with managers and employees.

Despite the initial smoothness of his entry into the organization, Tang encountered difficulties in fieldwork with regard to the "roles" in which he was cast by organization members. When he began his fieldwork in the organization, he was given the title of consultant by the director, in an effort to enhance his status in the eyes of the organization members—a title he accepted willingly until he discovered that it interfered with the research process. His difficulty was that people expected that he would inform them about the organization instead of talking about themselves. The title put him at the higher end of a hierarchical relationship; as consultant, he was the one expected to give answers. One informant said, "Mr. Tang, I don't think that my perspective is useful. I have had little education and I don't think my judgment is right at all. Since you are an expert with Western training, please don't be modest. I am eager to learn from you" [29, p. 304]. From then on, he sought a more modest introduction in order to shape a different kind of construction of his role.

Tang also found that his status as a U.S. resident cast suspicions on his purposes. In one incident, members became aware that he was taking fieldnotes in English. This led them to jokingly ask if Tang was working for the CIA. Recognizing the serious concern underlying the joke, Tang quickly apologized and promised not to write in English again.

It is likely that Tang's cultural familiarity enabled him to sense meanings underlying the encounter that a non-native researcher would overlook. What is most significant about the lack of trust expressed by the participants is that it alerted Tang to the importance of attending to "trust" (and lack of trust) as an organization-wide phenomenon. In this case, the difficulties of establishing trust in the research relationship created the possibility for a richer description of the organizational climate itself. Here again, it was the researcher's analysis of the research process, what Tang refers to as his "metaconversation" with the material that led to a rethinking of what counts as data.

CONCLUSIONS

In this chapter I have tried to problematize the issue of researcher "familiarity" as it arises in organizational fieldwork. In the case of the three cross-cultural research projects described here, a primary motivation emerged from the students' unique vantage points as native observers. Their cultural knowledge sensitized them to discrepancies between the particularities of the cultural setting and empirical generalizations about those settings in the literature.

In a variety of ways, the observer-observed relationship "delimits and defines the research process" [30, p. 49]. First, researcher access to the organization may be facilitated, or limited, by the members' construal of who the researcher is and what they think he or she is up to. Second, such interpretations will shape the nature of the data elicited; how research subjects respond to questions will depend in great measure on how they make sense of the researcher's purposes, and how they see those purposes as aligning with their own organizational goals. Finally, research *problems,* in the form of subjects' resistances and rejections, can be a source of relevant insight about the research topic, rather than simply bad data to be eliminated from the analysis.

The significance of these issues extends beyond the researching and writing of dissertations by foreign students, for the examples offered here also speak to the importance of reflexive inquiry in the everyday practices of communication professionals in intercultural settings. Communication professionals, by definition, serve a bridging function within and among organizations, spanning the boundaries of departments and larger organizational units in order to collect and disseminate information to facilitate the work of others. Successful bridging in intercultural settings requires an awareness of the different kinds of realities, the ways that individuals frame and shape their worlds, and what happens when people with similar and different ways of framing reality come together for work.

As an illustration of reflexivity in professional practice, I offer the following example from the domain of information systems design. It concerns the experiences of two organizational consultants who were engaged in the redesign of a communication system in a scientific organization [31]. This "lessons learned information system" was an on-line system for sharing technical lessons that have been learned by employees. It was originally designed as an electronic repository of records of "errors" in routine work practices, along with implications for the prevention of future mistakes; it was conceived as a kind of library that people could consult when they needed information about how to do their technical, day-to-day work.

This system, though seemingly sound from the designers' point of view, was not widely used for information sharing in the organization; potential users tended to dismiss it as simply a record of mistakes, a repository of "common sense" which offered no new insight into problems. Thus, the consultants were engaged in trying to change the organization's understanding of the system, from seeing it as a repository of records to a new means of learning through "dialogue" among users.

In their initial work with organization members to stimulate more creative and varied uses of the on-line system, the consultants encountered some resistance on the part of organization members [32]. The consultants eventually perceived that their point of entry into the organization was a constraining factor in the conversation. They had been invited by the organization's "Safety" specialists. Because the Safety division was perceived to concentrate solely on rules, regulations, and documentation, this identification worked against a conversation about creative opportunities for dialogue using the lessons-learned system. Only when this assumption was acknowledged and discussed within employee groups did a more exploratory conversation between consultants and users begin to take place.

Understanding the interpretive frames of prospective users or clients is a perspective-taking skill that pertains to the professional as well as the researcher. It requires that we become aware of our own activity in interaction, and of the influence that this can have on the activity of others. In any study of culturally distant contexts or of contemporary organizations, our identities as observers are socially constructed in the encounters with those we study. Recognizing and analyzing the effects of our memberships and frameworks is central to the investigative enterprise.

REFERENCES

1. H. Schwartzman, *Ethnography in Organizations,* Sage, Newbury Park, California, 1993.
2. B. Malinowski, *Argonauts of the Western Pacific,* Routledge, London, 1922.
3. F. Steier, *Research and Reflexivity,* Sage, London, 1991.
4. D. Berg and K. Smith, *Exploring Clinical Methods for Social Research,* Sage, Beverly Hills, California, 1985.
5. J. McLeod and J. A. Wilson, Corporate Cultural Studies and Anthropology: An Uneasy Synthesis, in *Anthropological Perspectives on Organizational Culture,* T. Hamada and W. Sibley (eds.), University Press of America, Lanham, Maryland, 1994.
6. M. Maruyama, Endogenous Research: Rationale, in *Human Inquiry: A Sourcebook of New Paradigm Research,* P. Reason and J. Rowan (eds.), John Wiley & Sons, Chichester, 1981.
7. T. Lindlof, *Qualitative Communication Research Methods,* Sage, Thousand Oaks, California, 1995.
8. M. Agar, The Intercultural Frame, *International Journal of Intercultural Relations, 18*:2, pp. 221-238, 1994.
9. E. Hall and M. Hall, *Understanding Cultural Differences,* Intercultural Press, Yarmouth, Massachusetts, 1990.
10. T. Hammond and A. Preston, Culture, Gender and Corporate Control: Japan as "Other," *Accounting, Organizations and Society, 17*:8, pp. 795-808, 1992.
11. G. Hofstede, *Culture's Consequences: International Differences in Work-Related Values,* Sage, Beverly Hills, California, 1980.
12. G. Hofstede, *Culture and Organizations: Software of the Mind,* McGraw-Hill, New York, 1991.
13. A. Kieser, Geert Hofstede: Cultures and Organizations, Software of the Mind, *Organization Studies, 15*:3, pp. 457-460, 1994.
14. J. Hunt, Applying American Behavioral Science: Some Cross-Cultural Problems, *Organizational Dynamics,* pp. 55-62, 1981.
15. S. Sackmann, *Cultural Knowledge in Organizations: Exploring the Collective Mind,* Sage, Newbury Park, California, 1991.

16. T. Hamada, *American Enterprise in Japan,* State University of New York Press, New York, 1991.
17. G. Bateson, Morale and National Character, in *Steps to an Ecology of Mind,* Ballantine, New York, 1972.
18. B. Pearce, Intercultural Communication and Multicultural Society: Implications for Communication Teaching and Research, *Teoria Sociologica,* 2:3, pp. 46-62, 1994.
19. R. I. Sutton, The Process of Organizational Death: Disbanding and Reconnecting, *Administrative Science Quarterly, 32*:4, pp. 542-569, 1987.
20. B. Kauffman, Feminist Facts: Interview Strategies and Political Subjects in Ethnography, *Communication Theory,* 2:3, pp. 187-206, 1992.
21. T. Ulrich, *Cross-Cultural Issues in the Management of German-Owned Business Subsidiaries in Hampton Roads,* unpublished thesis, Old Dominion University, Norfolk, Virginia, 1993.
22. T. Ulrich, personal communication, 1993.
23. G. Adams and V. Ingersoll, The Difficulty of Framing a Perspective on Organizational Culture, in *Organizational Culture,* P. Frost, L. Moore et al. (eds.), Sage, Beverly Hills, California, 1985.
24. A. Peshkin, In Search of Subjectivity—One's Own, *Educational Researcher, 17*:7, pp. 17-22, 1988.
25. M. L. Herweg, *Cross-Cultural Differences in Inter-Cultural Cooperation in the Context of Change and Uncertainty: Americans and Finns in the Workplace,* unpublished dissertation, Old Dominion University, Norfolk, Virginia, 1996.
26. M. L. Herweg, personal communication, 1996.
27. J. Lofland and L. Lofland, *Analyzing Social Settings,* Wadsworth, Belmont, California, 1984.
28. H. Chang and G. R. Holt, More than Relationship: Chinese Interaction and the Principle of Kuan-Hsi, *Communication Quarterly, 39*:3, pp. 251-271, 1991.
29. Y. Tang, *Cultural Transposition: Exploring Meanings of and Strategies for Cross-Cultural Transfer of Organizational Practices,* unpublished dissertation, Old Dominion University, Norfolk, Virginia, 1993.
30. C. A. Bailey, *A Guide to Field Research,* Pine Forge Press, Thousand Oaks, California, 1996.
31. F. Steier and E. Eisenberg, From Records to Relationships: Courting Organizational Dialogue at NASA, *Cybernetics & Human Knowing, 4*:1, 1997.
32. F. Steier, personal communication, 1997.

CHAPTER 14

Afterword: Teaching and Research Directions for International Professional Communication

DIXIE GOSWAMI

In his introduction, Carl Lovitt notes that this collection of chapters aims to help professional communication teachers prepare their students for the global workplace and asserts that "[t]he need to engage students and faculty in studying communication in the global workplace not only sets a new direction for research in this area but also anticipates fundamental changes in the way we teach professional communication—away from models of information transfer toward models of project-based learning and action research." Nowhere is the dynamic relationship among theory, research, and teaching more apparent than in these chapters, which taken as a whole call for an inquiry-based, constructivist pedagogy. By inquiry-based, I mean a program of preparation that immerses students in the experience and use of multiple discourses, in authentic simulations of intercultural relationships to be found in the global workplace. By constructivist, I mean teaching methods that set up projects or problems so that students can build knowledge for themselves through hands-on experience and problem solving rather than predominately through textbooks, exercises, and lectures. Another direction suggested from various perspectives in this volume is toward the teaching of analytical and heuristic skills, asking students to stand back from what they are studying to view it critically in relation to its intercultural context and to their own personal and group identity. Central to these pedagogies and practices is the issue of emerging technology that is transforming the academy, the global workplace, and the complex relationships that connect them: increasingly, "distance working," which includes everything from product development and documentation to the routine exchange of information by e-mail, is a defining feature of the global workplace.

The relatively open global communication environment and the flow of information across boundaries are realities that inform most of the studies reported here. Another reality, perhaps in contradiction to the open environment mentioned above, is that in spite of discourse about collaboration, culture, and shared values, the global workplace is market-driven. Students and specialists need to be able to speak up, critique the discourses of the workplace (and the classroom), notions of intercultural competence, and the conditions of their working lives.

This collection of chapters calls for the massive and challenging tasks of further developing the conceptual framework of the teaching of global professional communication and mapping this (evolving) agenda against existing teaching practices and textbook materials in order to extend teachers' repertoires. Obviously, this open-ended process will require multiple, cross-disciplinary, and intercultural collaborations and conscious, well-articulated links with research. Perhaps the most striking message is the call for transformed practice and theory-building that arises from interactive teaching and research.

Each chapter in this collection raises questions about teaching practices; about the links among theory, research, and pedagogy; about what students need to know; about what new discoveries and approaches might mean for continued growth of the discipline. Three studies in the volume illustrate the range and some of the major features of the teaching of global professional communication as it is represented here.

Tim Boswood, in "Redefining the Professional in International Professional Communication" develops a definition of professionalism in the context of global communication by means of nine propositions, seeing the professional as "an expert reconciler of difference, manager of intercultures, whose fundamental function is the creation of synthesis out of multiple discourses." Boswood asserts that the goal of educating students in the "building of a new, fluid identity" has practical educational consequences, two of which are echoed throughout this volume: teachers are themselves professional communicators, who must routinely analyze the discourse of the classroom and also prepare students to be able to analyze it; successful intercultural professional communication must be demonstrated and modeled in the classroom. Boswood argues that as teachers we must, on a daily basis, engage in analysis and reflection, inquiry and problem-solving, always improving our intercultural competence, if we are to be effective in the classroom. His view of teachers as active learners who are themselves professional communicators poses "daily personal and professional challenges for teachers equal to those faced by the students."

In "Intercultural Business Communication: An Interactive Approach," Jürgen Bolten states that static comparisons of different cultures, which tend to become formal "guides" for correct behavior in international business settings, miss the interaction process itself that decides the success or failure of communication strategies aimed at internationalization. He draws on Habermas and others to construct a "culture-sensitizing" model that focuses on individuals' store of life-knowledge as it affects intercultural communication in the rapidly changing realities of the global workplace. Bolten argues that "intercultural competence" is a general ability that can be developed in corporate training settings and in classrooms by means of simulations that initiate an intercultural process resembling reality so that students experience the simulation as an authentic situation. The aim is for students to react authentically in intercultural simulations and for the subsequent analysis of videotaped sessions to reveal very clearly the respects in which each individual's intercultural competence can be improved.

If intercultural competence—the ability to solve "intercultural problems" that arise when members of different cultures come into actual contact—is a necessary and basic skill for present and future communicators in the global workplace and if Bolten's theory of intercultural competence is valid, what is required? The simulations that lead to engagement in authentic intercultural sequences are dramatically different from the case studies that are routinely presented in professional communication textbooks: Bolten notes that both are important, but he claims that the active engagement of teams of learners in intercultural situations is necessary—in corporate settings and in classrooms—with the focus on global/multicultural interaction rather than on cultural boundaries. He further states that analysis, reflection, and transformed action are key parts of the developmental process leading to increased cultural sensitivity. His theory-driven approach relies heavily on problem solving and reflective practice.

Jane Perkins, in "Communicating in a Global, Multicultural Corporation: Other Metaphors and Strategies," argues that corporate theorists and interpretive ethnographers suggest new pedagogical directions for teachers of international professional communication. She states that teachers need to draw on "stories of the global workplace," understand the variety and pervasiveness of local contexts, and emphasize individuals and the disarray of cultures rather than "still lives" and boundaries if they are to develop relevant teaching theories and extend their repertoires of practices. Perkins and Bolten share the premise that teachers must also be theorists and researchers who have significant and continuous experiences in global workplace environments and first-hand understandings of the complex and pervasive nature of global communication. Boswood and other contributors develop and support this premise, by implication and directly.

What do teachers of global professional communication need? An agenda suggested by contributors might include:

- access to and participation in research that focuses on intercultural workplace and classroom discourses;
- knowledge of how to read and critique ethnographic narratives and empirical research reports;
- teaching strategies that demonstrate and model strategies for successful and ethical communication in the global workplace;
- the habit of critically examining methods, course materials, and practices for possible instances of cultural bias;
- high levels of intercultural competence and cultural sensitivity, as well as technical, contextual, and world knowledge;
- access to and familiarity with technology that is basic to global professional communication;
- continuous experience in global workplace environments and stable partnerships with global workplaces that provide sites for fieldwork and shared inquiry; and
- membership in small-scale networks that support radical changes in practice to meet the changing needs of students and encourage teachers in their roles as researchers and theorists.

We are advocating a teaching agenda that is not bounded and programmatic but that explores the range of variation important for meeting the changing needs of students of global professional communication, benefiting from powerful notions of intercultural communicative competence and a rich if fairly recent pedagogical history. The image of the teacher as researcher and theorist and the student as active, critical learner is central to the enterprise, as are cooperative arrangements between the academy and the global workplace.

Clearly, the teaching agenda generated by this volume invites dialogue and discussion. We can imagine forums where teachers, researchers, and workers critique the agenda presented here and design courses of study and programs of research that reflect their shared goals and understandings, resources and constraints, creating a lens through which current pedagogy and practice might be examined and plans for change considered.

In "Rethinking the Role of Culture in International Professional Communication," Carl Lovitt sums up the directions for research that arise from this volume with his call "to study the contexts in which such communication occurs and to analyze specific instances of workplace discourse," noting the shift from the national culture to situated practice in the global workplace. This statement describes the dynamic relationship between pedagogy and research implied by these specialists, suggesting that an important consequence of research should be to strengthen teaching by studying intercultural communication in the workplace—and in the classroom but not, I think, advocating an uncritical transference of workplace values and practices to pedagogy. Moreover, this is not to suggest that research in response to the needs of theory be neglected but that the direction should be toward applied studies conducted in the context of the global workplace and that intercultural, multidisciplinary research teams should include teachers, students, and specialists. Academia does not have a history of encouraging research that is responsive to practical and pedagogical issues. The corporate sector does not have a history of encouraging research that invites critique and negotiation. Movement in this direction will almost certainly generate political pressures, which in turn will focus attention on the integrity of research questions, design, and methods and other issues.

We have gathered readings that illustrate the complicated and divergent nature of intercultural communication studies. While acknowledging the limitations of any one perspective, we believe that this collection of essays nevertheless suggests some promising directions for future research, including:

- building on key studies of writing and speaking in non-academic settings—empirical and naturalistic—thoroughly analyzed and critiqued;
- conducting historical and anthropological investigations of the functions and uses of written and spoken language in intercultural workplace settings, including cyberspace;
- studying the ways in which novices achieve intercultural communicative competence, how they and experts use it, and the consequences of its achievement for individuals and organizations;
- investigating the social setting of intercultural discourse and its context in the classroom, the workplace, and electronically;
- designing research from various disciplinary perspectives that might be applied to the process of internationalizing document design process models; and
- preparing ethnographies and case studies to uncover situated practice, including detailed descriptive studies that show how speaking and writing integrate with managerial, technical tasks, and other workplace tasks.

I want to emphasize this point: from the perspective of this collection of chapters, an important task for teachers and researchers is to make the medium of intercultural discourse the object of focal attention as they investigate the multidimensional aspects of written and spoken language use in intercultural workplace settings. From this point, a host of ethical questions arise about research methodology, representation, and the centrality of interpretation to the enterprise.

Building an intercultural, multidisciplinary community of researchers, including teachers, students, and specialists, must be a goal as well as a consequence of any present and future teaching agenda and for research directions for global professional communication, notwithstanding the constraints that make this kind of cooperative effort challenging and difficult. Clearly, devising strategies that will make it possible for academia and the international business community to support inquiries that attend to the needs of theory and practice while maintaining intellectual integrity must be a topic of ongoing debate and dialogue.

Contributors

DEBORAH C. ANDREWS is Professor of English and Coordinator of the Business and Technical Writing Program at the University of Delaware, where she teaches undergraduate and graduate writing courses and conducts research on communication patterns in multinational organizations. She has been a writing consultant for several organizations, including AT&T Technologies, the American Chemical Society, Dominion Textile Inc., General Electric, Hercules Inc., and the National Science Foundation. She has published many scholarly articles in such journals as *Engineering Education, Technical Communication,* and the *Journal of Technical Writing and Communication.* She has also contributed chapters to several anthologies and co-authored five textbooks. She edited an anthology, *International Dimensions of Technical Communication,* published by the Society for Technical Communication Press in 1996 and has a new text, *Technical Communication in the Global Community,* published by Prentice Hall in November 1997.

STEPHEN A. BERNHARDT is Professor of English at New Mexico State University, specializing in technical and professional communication. He is also Senior Consultant, Scientific Services, Franklin Covey Company, Salt Lake City, Utah. Dr. Bernhardt teaches courses in workplace communication, training, and writing technologies. His book *Writing at Work: Professional Writing Skills for People on the Job* was published by NTC-Tribune Publishing Group (Lincolnwood, Illinois) in 1997. He is President of the Council for Programs in Technical and Scientific Communication and Vice President of the Association of Teachers of Technical Writing. His consulting and training work involves high tech computer companies, hospitals, and pharmaceutical companies.

JÜRGEN BOLTEN, Professor of Intercultural Business Communication at the Friedrich Schiller University in Jena, Germany, teaches courses in intercultural business communication and conducts research on the theory of intercultural business communication, intercultural management, cultural anthropology, and business languages. He also works as a consultant in the fields of international marketing and management. Formerly, he founded and directed the Institute for International Communication at the Heinrich Heine University in Düsseldorf. He has edited a number of books on business languages and intercultural communication, most recently the two volumes *Cross Culture—Interkulturelles Handeln in der Wirtschaft* (*Cross Culture: Intercutural Behavior in Business*) and *Transformation und Integration. West-/osteuropäische Wirtschaftsbeziehungen* (*Transformation and Integration. West/East European Economic Relations*), both published in 1995. His publications also include textbooks for Business German, such as *Marktchance Wirtschaftsdeutsch* (Business German for advanced learners of German as a Foreign Language), 1993, and several scholarly articles. He is currently writing another textbook for Business German, and a textbook for Intercultural Negotiation Training is about to be published.

DEBORAH S. BOSLEY is Associate Professor of English, Internship Coordinator, and Program Advisor for the Technical and Professional Writing Program at the University of North Carolina at Charlotte, where she teaches courses in technical communication at the undergraduate and graduate levels. For the past fifteen years, she has been a technical communications' consultant to such companies as Royal Insurance, Hoechst Celanese, Illinois Power Company, KPMG Peat Marwick, IBM, First Union National Bank, and the Mecklenburg County Department of Social Services. She has published articles in *Technical Communication Quarterly, Technical Communication, the ABC Bulletin Quarterly,* and *IEEE Transactions on Professional Communication.* In 1992, she received an award for Outstanding Article from STC and in 1993, she received the Nell Ann Pickett Award for Article of the Year presented by ATTW. She is the former President of the Metrolina Chapter of STC and has held positions on the NCTE Commission on Technical and Scientific Communication and the Committee on Instructional Technology. Her research interests include collaboration and teamwork, particularly cross-cultural collaboration, and international graphics.

TIMOTHY S. BOSWOOD is Associate Professor of English at the City University of Hong Kong, where he teaches courses in organizational communication, technical writing, and document design. His current research focuses on identity negotiation in management writing. He has co-edited two books on English for professional communication, and his articles have appeared in the *TESOL Journal* and in *English for Special Purposes.*

JUDI BROWNELL is Professor of Managerial Communication and the Richard J. and Monene P. Bradley Director for Graduate Studies at the School of Hotel Administration, Cornell University, where she teaches undergraduate and graduate courses in organizational and managerial communication and participates regularly in executive education programs. She has designed and conducted training seminars for a wide range of hospitality, educational, and other work organizations. Dr. Brownell is the author of several business and communication texts, including *Organizational Communication and Behavior: Communicating for Improved Performance* (Holt, Rinehart, and Winston, 1989) and *Listening:*

Principles, Attitudes, and Skills (Allyn and Bacon, 1996). She has published over fifty articles and book chapters. Professor Brownell serves on the editorial boards of five professional journals, has won several research awards, and is past president of the International Listening Association. She is past president of the New York State Communication Association and is past president of the Southern Tier Chapter, American Society for Training and Development.

LINDA DRISKILL is Professor of English and Administrative Science at Rice University, where she administers the university's Writing Program and teaches undergraduate courses in managerial communication and engineering communication and a graduate seminar in contemporary approaches to teaching composition. She co-authored *Business and Managerial Communication: New Perspectives* (Harcourt Brace Jovanovich, 1992) and *Decisive Writing* (Oxford University Press, 1978). She has also published articles in business and technical communication. Her most recent article, "Writing Across Borders: Managing Collaboration and Changing Our Perspectives," focuses on changing practices of collaborative writing in multinational corporations. She has consulted for corporations in engineering and finance, including Stanford Financial Group, Rotan Mosle, Criterion Investments, and Vista Chemical Company, and with government agencies in the design of technical manuals and reports.

DIXIE GOSWAMI is a Senior Scholar for the Strom Thurmond Institute of Government and Public Affairs at Clemson University, where she directs a national company-based literacy program. As a former Professor of English at Clemson University, she taught undergraduate and graduate courses in professional communication, international business communication, and composition theory. In 1989 she received the University Award for Excellence in Teaching, Scholarship, and Community Service. The recipient of a U.S. Department of Education Mina Shaughnessy Fellowship, Professor Goswami is a former Blumenthal Scholar in Writing at the University of North Carolina, Charlotte, and Visiting Professor of English at the University of Massachusetts, Boston. She was recently named a Visiting Fellow at Brown University by the Annenberg Institute. She also directs the program in writing at the Bread Loaf School of English, Middlebury College, and she coordinates the Bread Loaf Rural Teacher Network. She has published numerous articles on writing, teaching, and learning, and co-edited several books, including *Writing in Nonacademic Settings* (with Lee Odell). As a research scientist with the American Institutes for Research, she co-authored AIR's *Writing in the Professions.*

JANE JORGENSON is Assistant Professor in the Department of Communication at the University of South Florida. She has also taught in the Department of Engineering Management at Old Dominion University, where she also held a position as Research Associate in the Center for Learning Technologies. Dr. Jorgenson has taught professional communication via television as part of Old Dominion's statewide Teletechnet system, as well as through National Technological University, which broadcasts graduate courses to corporate sites nationally. She has published book chapters and scholarly articles in journals such as *The Journal of Applied Communication Research, Communication Theory,* and *Teoria Sociologica.*

MICHAEL KEENE is Professor of English at the University of Tennessee, Knoxville, where he created and is former director of the concentration in technical communication and teaches in the graduate program in Rhetoric and Composition. He has worked as a consultant in writing and editing for Martin-Marietta Energy Systems at Oak Ridge National Laboratory, Lockwood-Greene Engineering, JBF Engineering, the University of Tennessee, Tombras Advertising Agency, and Eastman Chemical Company. His publications include these books: Mayfield's *Quick View Guide to the Internet for Students of English* (Mayfield, 1997, with Jennifer Campbell); *The Easy Access Handbook* (Mayfield, 1996, with Kate Adams); *The Heath Guide to College Writing* (D. C. Heath, 1992, 1995, with Ralph Voss); *A Short Guide to Business Writing* (Prentice-Hall, 1994, with Harry Bruce and Russel Hirst); *Effective Professional and Technical Writing* (D. C. Heath, 1987, 1992); and the revised Eighth Edition of W. Paul Jones's *Writing Scientific Papers and Reports* (Wm. C. Brown, 1980). He has also published numerous articles on composition and technical communication in such journals as *Technical Communication, Journal of Advanced Composition,* and *College English,* as well as chapters in *Teaching Advanced Composition: Why and How* (Boynton/Cook, Heinemann, 1991); *Establishing and Maintaining Internships* (ATTW, 1989); *Perspectives on Research and Scholarship in Composition* (MLA, 1985); *Research in Technical Communication* (Greenwood, 1985); - *Teaching Business, Technical, and Scientific Writing in the Two-Year College* (NCTE, 1983); and *Teaching Audience Analysis and Adaptation* (ATTW, 1980). His current book-length projects include a profile of ten different types of technical communication programs, *Education in Scientific and Technical Communication: Programs That Work* (STC, 1997, with Freda Stohrer, Sam Geonetta, Russel Hirst, and others), funded by a major STC research grant, and *The Work of Happiness,* a multicultural reader co-edited with Marilyn Kallet, for Harbrace.

CARL R. LOVITT is Associate Professor of English at Clemson University, where he teaches undergraduate and graduate courses in international professional communication. A former administrator for the Modern Language Association, he is also Director of the Pearce Center for Professional Communication, a $3-million endowed center that sponsors a wide range of programs to improve teaching and research in all areas of communication. His classroom-based research recently involved having teams of students conduct on-site ethnographic studies of communication practices in such multinational corporations at BMW, Hitachi, Michelin, and Ryobi. He co-directed the first three National Writing Across the Curriculum Conferences. Dr. Lovitt's translations and articles have appeared in numerous books and journals, and he currently has a composition textbook under contract to Addison Wesley Longman. He is a frequent speaker and workshop teacher at professional conferences, and he currently serves on the International Committee of the Association for Business Communication.

JANE M. PERKINS is Associate Professor of English at Clemson University, where she teaches professional communication courses and conducts workplace communication research. Drawing on her three-year ethnographic study of communication in an international software development corporation, she has published and presented papers on the relationship between intercultural communication and technology transfer. Her articles have appeared in *Studies in Technical Communication, IEEE Transactions on Professional Communication,* and *The Bulletin of the Association for Business Communication.* She is a frequent presenter at international and national conferences, including the Association for Business Communication, Computers and Writing, Conference on College Composition and Communication, Modern Language Association, and Society for Technical Communication.

FREDERICK STEIER is Director of Interdisciplinary Studies and Professor of Communication at the University of South Florida. A former Associate Professor of Engineering Management and Executive Director of the Center for Cybernetic Studies in Complex Systems at Old Dominion University in Norfolk, Virginia, he has served as President of the American Society for Cybernetics. He was the King Olav V Fellow at the University of Oslo, where he investigated cross-cultural communication issues in diverse organizations and therapeutic settings. He has lectured and worked extensively with organizations on issues of intercultural and organizational communication and design throughout North America and Europe. A current project with NASA deals with the creation of an organizational dialogue around sharing lessons learned from professional cultures. The editor of *Research and Reflexivity* (1991), his research has been published in journals such as *The Journal of Applied Communication Research, Teoria Sociologica,* and *The Journal of Communication.*

ELIZABETH TEBEAUX is Professor of English at Texas A&M University and Professor of Managerial Studies at Rice University. She served as Coordinator of Technical Writing at Texas A&M for fifteen years and is now launching a business communication program for the College of Business at Texas A&M. She is author/co-author of three books on business and technical communication: *Writing Communications in Business and Industry* (Prentice-Hall, 1982), *Design of Business Communication* (Macmillan, 1990), and *Reporting Technical Information* (8th edition, Macmillan, 1994). In addition, she has authored seventy articles on technical communication, one of which won the NCTE award for best article on technical communication pedagogy in 1989. Dr. Tebeaux has served as President of the Association of Technical Writing, and she was a founding member of the CCCC Committee on Technical Communication, which she chaired for three years.

S. PAUL VERLUYTEN is a Professor of Theoretical Linguistics at the University of Antwerp, where he teaches courses in business communication, intercultural communication, and business French. He has taught courses on intercultural communication in business at Ateneo University, Manila, Philippines; at the Graduate School of Business Administration of the National Institute of Development Administration, Bangkok, Thailand; and at the Center for European Studies, Maastricht, Netherlands. He also teaches seminars on intercultural communication for the headquarters staff of NATO in Brussels and for the staff of SABENA Airlines and serves as a trainer and consultant on intercultural issues for various companies and institutions in Belgium. He is the author of several articles on intercultural communication, and he is currently preparing a book on intercultural communication in the world of business and institutions.

JOHN WEBB is a General Partner in Concurrent Communications, a technical communication consulting and contracting firm in Knoxville, Tennessee. His clients include the Electrical Power Research Institute and the Department of Energy. He is also the managing editor of the *Pollution Prevention Advisor,* a quarterly national newsletter of the U.S. Department of Energy. His publications include articles in *Pollution Prevention Review, D&D Technologies,* and *TIE Quarterly.* A book chapter co-authored with Michael Keene will appear in the forthcoming Ablex volume, *Knowledge Diffusion in the U.S. Aircraft Industry.* He is currently pursuing a graduate degree in technical communication at the University of Tennessee, Knoxville.

TIMOTHY WEISS is an Associate Professor at The Chinese University of Hong Kong. He has taught at the University of Kansas, Iowa State University, the University of Illinois at Urbana-Champaign, and the University of Maine. On two occasions he has been a Senior Fulbright Scholar: in Tunisia (1988-89) and in Algeria and Morocco (1993-94). He has also served as a consultant in Expository and Professional Writing to the Council for International Exchange of Scholars in Washington, D.C. He has published articles on professional communication and on literary topics; he is author of a monograph, *Fairy Tales and Romance in the Works of Ford Madox Ford* (1984), and a book of literary criticism, *On the Margins: The Art of Exile in V.S. Naipaul* (University of Massachusetts Press, 1992). His current research interests include international communication and orientalism.

Index